A Bayesian Model Framework to Determine Patient Compliance in Glaucoma Cases

A Bayesian Model Framework to Determine Patient Compliance in Glaucoma Cases

Raghu Babu Korrapati, Ph.D

iUniverse, Inc.
New York Lincoln Shanghai

A Bayesian Model Framework to Determine Patient Compliance in Glaucoma Cases

iUniverse books may be ordered through booksellers or by contacting:

iUniverse
2021 Pine Lake Road, Suite 100
Lincoln, NE 68512
www.iuniverse.com
1-800-Authors (1-800-288-4677)

ISBN-13: 978-0-595-36839-6 (pbk)
ISBN-13: 978-0-595-81252-3 (ebk)
ISBN-10: 0-595-36839-5 (pbk)
ISBN-10: 0-595-81252-X (ebk)

Printed in the United States of America

For my son

Surya Korrapati

Contents

List of Illustrations

List of Tables

Executive Summary

A Bayesian Model Framework to Determine Patient Compliance in Glaucoma Cases

By

Raghu Babu Korrapati, Ph.D

The purpose of this research was to devise a Bayesian model framework to assess the compliance with medication in glaucoma patients. Bayesian Networks have increasingly become tools of choice in solving problems involving uncertainty in the medical domain. These models have been successfully applied to diagnosis applications. This research applied Bayesian modeling to medication noncompliance in glaucoma patients. Medication noncompliance is the failure to comply with a physician's instructions with regard to taking medications at specified times. If the patient is non-compliant, irrespective of the advances in medical field, the person does not benefit from medical intervention. A model-based decision support system using a Bayesian Network was developed to determine whether a patient was complying with the medications prescribed by the physician. The predictive ability of the model was investigated using the existing patient data. To assess research validity, the results obtained through the model were compared against a domain expert's evaluation of the patient cases. The results provided by the Bayesian framework agree with the information provided by the domain expert. The Bayesian model can be used to confirm an ophthalmologist's clinical intuition or to formulate a prescription strategy for a glaucoma patient. The model can be refined using larger patient data sets and additional variables can refine the model. A clinical decision support system can be developed using the refined model to prevent medical errors in glaucoma compliance process. Results from this study could potentially improve the decision making process, given the uncertain and incomplete data available to a physician. The model may be generalized to other business situations where a decision has to be made based on incomplete and uncertain datasets.

Acknowledgements

I would like to express my sincere gratitude to Dr. Sumitra Mukherjee, without whose mentoring this research would not have been possible. I am indebted to him for his constant encouragement and support that made this research a success. I extend gratitude to Dr. K. V. Chalam, who has been instrumental in this research and especially for making data available for this research. I thank Dr. Michael Laszlo for making suggestions that made this research the best it could be.

I owe Dr. Marco G. Valtorta special thanks for guiding me through the Bayesian Networks procedures used in this research and for permission to use the Artificial Intelligence Lab at the University of South Carolina. Special thanks to Dr. Moninder Singh of the IBM Thomas Watson Research Labs for providing the EM program and instruction in working with the current release of Hugin Lite. I appreciate his patience in answering all of my questions regarding the use EM algorithm in this research. I thank Dr. Bijoy Sahoo for his direction and constant moral support in publishing the initial versions of this research, as well as other papers, at the Allied Academies National Conference in the Academy of Information and Management Science and its research journal.

Finally, thanks to my wife, Vijaya, for her patience and understanding during the entire process of my doctoral program. Her loyalty, moral and emotional support has helped me a long way throughout my life. I also extend sincere thanks to my brothers and friends for their constant support and encouragement during this time. Of course, I owe sincere thanks to the Lord Almighty, without whose grace and blessings this effort would not have been possible.

1

Introduction

Statement of the Problem to be Investigated and Goals to be Achieved

The purpose of this research was to devise a Bayesian framework to assess compliance with medication in glaucoma patients. A model-based decision support system using Bayesian Belief Networks (BBNs) was developed to determine whether a patient is complying with the medications prescribed by the physician. Glaucoma was selected because of the inherent uncertainty in determining medication compliance in the treatment process.

Specifically, this research aimed to achieve the following:

1. Identify potential variables affecting patient compliance behavior in glaucoma cases and investigate the inter-relationships among such variables.

2. Model the patient compliance problem and the strategy used by ophthalmologists as a Bayesian Belief Network (BBN).

3. Estimate subjective probabilities to represent the interdependencies among the variables of interest in the BBN.

4. Use an appropriate algorithm to infer posterior probabilities for events of interest, given a set of evidence.

5. Test and refine the BBN to verify the appropriateness of the model to predict compliance behavior with reasonable accuracy.

Relevance and Significance

Glaucoma is a clinical condition characterized by increased intraocular pressure (IOP) in the eye that results in damage to the optic nerve, loss of visual field, and eventual blindness. It predominantly affects people over the age of sixty. IOP is controlled with several medications that must be administered at regular intervals for optimum control (Cooper, 1996). A consistently high IOP that is not controlled with adequate medication can permanently damage the vision. The progressive damage is measured by two criteria: (1) visual field loss (the area one can see around a fixed spot) and; (2) changes in the optic nerve cup.

The patient is examined at regular intervals at the clinic where the doctor examines the IOP and changes in the visual field and optic nerve cup. After each examination, the physician potentially alters the plan of treatment according to the therapeutic guidelines based on available data. This is a standard procedure adopted by practicing ophthalmologists (Gaebel, 1997).

As Glaucoma affects mostly the elderly, compliance (usage of drugs at scheduled intervals as prescribed by a physician) is a major problem (Patel and Spaeth, 1995). At present, patient compliance cannot be predicted systematically (Duane and Jaeger, 1988). A physician's intervention is best guided by subjective estimates of patient compliance (Cramer, 1998). This method is potentially liable to human errors of judgment, but can be improved by using models such as the one developed in this research.

Because of the chronic nature of glaucoma, compliance with medication is very important. Compliance encompasses not only the frequency of medication but also patient awareness and interest in health, keeping appointments and making appropriate lifestyle changes. Noncompliance can range from accidental lapses (missing medication at random) to premature cessation of medication or ineffective and incorrect methods of taking medication (Murphy and Coster, 1997; Rotchford and Murphy, 1998). If the patient is noncompliant, irrespective of advances in the medical field, the health of the patient does not improve (Bloch, Rosenthal, Friedman, and Caldarolla, 1997).

Studies indicate that cases of noncompliance can be caused by one or more of the following (Brown, Brown, and Spaeth, 1984):

- Improper application of the drops in the eye(s)

- A missed dosage, accidentally or otherwise

- Not maintaining the sterility of the bottle
- Not washing the hands before applying the medication
- Any other form of noncompliance not mentioned above

In addition, studies have indicated that the patient's inability to understand the correlation between glaucoma and loss of vision and other psychosocial factors have a significant effect on the likelihood of compliance (Chang, Lee, Petursson, Spaeth, Zimmerman, Hoskins, Mills, Brown, Kass, and Lue, 1991; Kosoko, Quigley, Vitale, Enger, Kerrigan, Tielsch, 1998).

In this study, a model was developed to analyze and assess noncompliance behavioral factors in glaucoma, considering data that indicates the worsening of glaucoma, juxtaposed to identify cause-and-effect relationships among the variables. The model ideally benefits the ophthalmologist with more realistic estimates of the duration of noncompliance and elevated or unhealthy IOP. Apart from identifying patients with noncompliance and elevated IOP, the model may be valuable in understanding the rate at which the disease worsens due to noncompliance, duration of elevated IOP, or both. In this research, it was assumed that unhealthy IOP is synonymous with elevated or increased IOP.

Barriers and Issues

One of the major barriers to developing a decision support system to assess medication compliance in glaucoma cases is the characteristic degree of uncertainty involved in the overall process. This uncertainty could be due to several factors:

- Imperfect understanding of the domain
- Incomplete knowledge of the state of the domain at the time a given task is performed
- Randomness inherent in the mechanisms governing the behavior of the domain
- A combination of any and all of the above factors

Physicians often rely upon past experience and the available data to assess the likelihood of medication compliance. Several variables impact the assessment process where the relationship between these variables and their varying impacts gives rise to a subjective probability distribution. Thus, the key research issues that developed were:

- To model the degree of uncertainty involved in the medication compliance process,

- To model the physician's experience in terms of subjective probabilities, and

- To develop a technique that is useful in reasoning in this domain.

The probability that each patient complied with the medication prescribed can be best determined by relying upon available evidence. Bayesian inference has been used successfully in medical diagnosis systems for decades (de Dombal, Leaper, Staniland, Horrocks, and McCann, 1972). In this study, Bayesian inference was used to establish various decision rules and modify initial evidence, in light of new information.

A Bayesian approach is appropriate in assessing medication compliance among glaucoma patients because it:

- relies on a well-established classical probability theory,

- can handle problems when some data entries are missing,

- uses a global interaction between variables of interest,

- is one of the best representations available that combines prior knowledge and data,

- has the ability to predict results of interventions

The probability for some of the variables was ascertained with the help of an ophthalmologist, whose expertise comes from medical training, licensure, experience, and continuing education. The prior or known probabilities are a subjective interpretation of these variables by the domain expert. The prior probability values were refined by the clinical data used in this research.

Hugin Lite was used to develop the probability model and refine the probabilities. The Hugin system is a popular tool in constructing model-based decision support systems in domains characterized by inherent uncertainty. The models supported are Bayesian Belief Networks and influence diagrams. The refined model was used to predict the compliance of glaucoma patients based on the future clinical data.

Research Questions Investigated

The key research questions this study investigated were:

1. What are the potential variables affecting the medication compliance process in the glaucoma domain?

2. How can the degree of uncertainty involved in medication compliance process be modeled?

3. How can the domain expert's experience, in terms of subjective probabilities, be captured?

4. What are the techniques that are useful in reasoning in this domain?

Research Goals

The goal of this research was to build a model-based decision support system using Bayesian Belief Networks (BBNs) to determine whether a patient was complying with the medications prescribed by the physician. Specifically, this research identified variables for patient compliance behavior, developed a BBN model, and made inferences using probability values.

Limitations and Delimitations of the Study

The limitations of this study were as follows:

1. The duration of the uncontrolled IOP and noncompliance of the patient needs to be refined with experimental data and appropriate statistical analyses. Results based on empirical data may be inadequate given the available limited data set, limiting the predictive value of the research effort and the attendant generalizations.

2. The study includes a restricted number of variables to identify a simple methodology using Bayesian Belief Networks to ascertain the approximate noncompliance and uncontrolled IOP. The included variables, in their simplicity, may miss the complexity that surrounds both the onset and the manifestation of glaucoma in affected patients.

3. More variables can be included to ascertain their significance in
    determining the noncompliance, if this study is successful. Thus,
    even with the recognized limitations, this research provided a good
    starting point and the potential to positively contribute to the existing
    literature and expand the relevant research horizon.

Definition of Terms

Glaucoma

Glaucoma is caused by a number of different eye diseases that result in
increased pressure within the eye. The fluid backup in the eye causes elevated
pressures and damages the optic nerve over time. With early detection, diagnosis, and treatment, permanent loss of vision can be prevented.

The aqueous humor, produced by the ciliary body situated behind the iris,
constantly circulates through the anterior chamber. The aqueous humor flows
between the iris and the lens, nourishes the cornea and lens, and then flows
out through a very tiny, spongy tissue called the trabecular meshwork. The
trabecular mesh serves as a drain for the eye. When this drain becomes
clogged, aqueous humor cannot leave the eye, causing the fluid to back up.
But since the eye is a closed compartment, the backed up fluid causes
increased pressure to build up within the eye. This form of glaucoma is called
open (wide) angle glaucoma.

Intraocular Pressure (IOP)

The pressure inside the eye—the intraocular pressure (IOP)—builds up when
the aqueous humor is prevented from draining properly. The increase in pressure can damage the optic nerve and result in the loss of vision. Elevated IOP
is a major risk factor for glaucoma, and studies have indicated that it can sig-
nificantly damage the eye.

Optic Nerve Cup

The optic nerve of the eye carries visual information to the brain. It is made up
of over one million long, thin nerve cells. When the pressure in the eye builds
and exceeds the perfusion pressure, the nerve cells are compressed and sub-

jected to damage. In extreme cases, this condition can lead to the death of nerve cells, resulting in permanent visual loss. Early detection and treatment of glaucoma can prevent such loss.

Unhealthy IOP changes the shape of the optic nerve to form a cup. Loss of nerve tissue (an integral part of the optic nerve) leads to increased cup size. If glaucoma is left untreated and the damaged optic nerve cup enlarges, irreversible and permanent loss of vision can result due to prolonged elevated IOP.

Visual Field Test

The degree of visual field loss can be measured on a regular basis via a special apparatus. Morphological changes in the volume of optic nerve cup can be observed clinically at the time of patient visits. This test allows the physician to know if and how the patient's field of vision has been affected by glaucoma. The visual field is an important measure of the extent of damage to the optic nerve. The visual field test is conducted with a computer or with the Goldmann perimeter, both of which test the peripheral vision.

Summary

Glaucoma is a progressive, detrimental clinical condition that is prevalent among the elderly. Accurately predicting noncompliance with medication in these patients has been a challenge to the medical profession. In this study, a Bayesian Belief Network was used for modeling problems, characterized by inherent uncertainty of predicting noncompliance and the duration of the elevated IOP.

The known probabilities associated with the variables in the model were assigned, using an experienced ophthalmologist. The model was tested using clinical data which was refined and used for predicting noncompliance and the duration of elevated IOP with reasonable accuracy. The premise of this research was that if the model was found to be appropriate in predicting noncompliance and the duration of elevated IOP among glaucoma patients with reasonable accuracy, it could be used with appropriate modifications and refinements for other diseases with inherent uncertainties.

The succeeding chapters are organized as follows. Chapter 2 presents an overview of the existing research and the different methods available to solve the problem. In this chapter, the benefits of using Bayesian networks over the other competing methodologies are explained. The chapter also discusses the benefits of this research to the medical domain.

Chapter 3 discusses implementation methodology and the techniques that were used to solve the present problem. The chapter further explains the variables used in the modeling in which the prior probabilities of the known variables are enumerated. A sample case is worked out using the Hugin Lite system, highlighting the features of this tool. In addition, the format for presenting the results and the projected outcomes are discussed. The chapter concludes with an explanation of resource requirements and predictions of the reliability and validity of data/methodology.

Chapter 4 discusses the results of this study and contains the description of the implementation of Expectation-Maximization (EM) algorithm and the actual outcomes, including the data analysis. Constraints and limitations of this study are also considered.

Chapter 5 concludes the study and presents implications, recommendations, and a summary of the research. Additionally, recommendations for future research, considering additional affective factors, are also presented. The chapter assesses both the contribution this study makes to the field of Bayesian inference in medicine and its concurrent benefit to relevant academic research.

2

Review of Literature

In this chapter, a review of relevant research in the field of modeling and the associated tools to model the problem of noncompliance is presented. Different methodologies for developing a model are considered for noncompliance among glaucoma patients. This section of the research maps the research frontier and places the present research efforts within the context. The chapter concludes with a section describing the contribution that this research makes to the glaucoma research domain.

Historical Overview of Theory and Research

The reasons for using a Bayesian Network in modeling the problem, and its advantages over other competing methodologies are explained in the following sections. Two competing methodologies are Expert Systems and Artificial Neural Networks. The inadequacy of these competing techniques and their hollowness for investigating the current research problem are also discussed.

Theory and Research Literature Specific to the Topic

Expert Systems

Expert Systems (ESs) provide one of the earlier attempts to model an expert's reasoning. A rule-based Expert System consists of a set of production rules in the form of: IF (condition) THEN (fact or action). In real world situations, these rules are not absolutely certain as the information gathered is typically subject to uncertainty. These uncertainties are modeled using certainty factors associated with the facts and rules. Often these systems draw incorrect conclusions, since certainty factor calculus is not consistent and a combination of several uncertainty rules is not a local phenomenon, but rather, fully dependent on the entire (global) problem situation. Further, these systems may not be successful because of the intricate mathematical calculations involved in the process and their intractable nature (Gorry, 1973). Expert Systems are not useful in determining medication compliance because of the inherent difficulty involved in processing uncertainty at a global level.

Artificial Neural Networks

Artificial Neural Networks (ANNs) consist of several layers of nodes: primarily, a layer of input nodes and a layer of output nodes. In between the input and output nodes, a few hidden layers are imbedded. ANNs are successful in the areas of pattern recognition, but their drawback lies in the amount of time involved in training the network. Hence, ANNs can handle relations that are uncertain only with proper training of the network, but cannot help predict unforeseen situations when there is insufficient data to train it. Because of their inability to read the uncertainty of the conclusion from the network, ANNs may not be useful in determining medication compliance. This may be due to the unknown operations in the hidden layers of the network.

Wiegerinck, Kappen, Neijt, Van Dam, Braak, and Burg (1999) showed that modeling the problem of anemia was difficult using neural networks and the training of the system was very tedious. Physicians were not impressed with the system, but were more interested in one that could identify rare cases. Hence, the neural network project used to model the problem was abandoned in favor of Bayesian networks. Although the probabilistic rule-based system was useful in modeling the problem and able to handle the rare cases with relative ease, results showed that a detailed and very accurate diagnosis with Bayesian networks was not only possible, but had greater advantages over neural networks.

Compared to Artificial Neural Networks, Bayesian networks are advantageous because (1) the domain expert can provide knowledge in the form of causal network structures; (2) the network structures are understandable and extensible, and, (3) the networks can be used easily with missing or incomplete data.

Bayesian Networks

A Bayesian Network (BN) is an acyclic graph with links that organize the body of knowledge in any given area, by mapping the cause-and-effect relationships among the variables of interest. These cause-effect relationships are normally not completely deterministic as in rule-based Expert Systems. The strengths of these relationships are modeled with probabilities. BN uses a Bayesian probability theory and models the dependencies from structural descriptions of the problem.

Bayesian Networks for Representing Time Series

Bayesian Networks offer a powerful framework to solve problems in domains where dependencies among variables are known. Because of the inherent characteristic of this framework to deconstruct problems into several manageable subsystems, this approach also has been used in dynamic modeling (Andreassen, 1994; Berzuini, Bellazi, Quaglini, and Speigelhalter, 1992). Riva and Bellazi (1996) constructed a method to perform learning of probabilistic networks representing time series through model selection. Through this research, it is better understood that BN not only deals with static knowledge but also is effectively used to represent non-linear stochastic input-output models.

Temporal Abstractions in the Medical Domain

Temporal Abstractions (TAs) is an Artificial Intelligence technique that can be used to analyze and interpret longitudinal data. In the medical domain, TA mechanisms provide an abstract description of the patient's state at a given

time. Hence it is useful in understanding a patient's response to current proto-
col and provides proper therapeutic suggestions (Larizza, Bellazzi and Riva,
1997). The contribution of this research shows that Temporal Abstractions
can be combined with classic statistical techniques—such as time series analysis—in developing decision support systems for long-term monitoring of
chronic patients.

Bayesian Models in the Medical Domain

There are several Bayesian network systems developed for the medical diagnosis domain (Andreassen, Woldbye, Falk, and Andersen, 1987; Heckerman
and Nathwani, 1992; Heckerman, Horvitz, and Nathwani, 1992; Korver and
Lucas, 1993). The most well known of these is the MUNIN, a system for the
interpretation of electromyographic findings for the diagnosis of muscle and
nerve diseases. Some existing systems use probabilistic networks for treatment
planning in medicine (Andreassen, Hovorka, Benn, and Olesen, and Carson,
1991; Bellazi, Berzuini, Quaglini Spiegelhalter, and Leaning, 1991; Quaglini,
Bellazi, Stefanelli, and Locatelli, 1993).

Andreassen et al (1991) utilized probabilistic networks in their Diabetic
Advisory System that simulated the probability distribution for blood glucose
concentration over a period of time. In relation to this work, Oppel, Hierle,
Janke, and Moser (1993) reported and represented compartmental models,
through stochastification using probabilistic networks. A similar approach was
also used to build a model-based patient monitoring system (Berzuini et al.,
1992).

Kahn, Roberts, Wang, Jenks, and Haddawy (1995) used Bayesian Networks in the study of breast cancer for radiological clinical decision support. A
model based on the interpretation of the mammogram was developed from a
limited number of patient histories. In this study, a set of 77 cases from a
mammography atlas were analyzed using MammoNet to determine whether
lesions were benign or malignant. The same data was analyzed using a Bayesian Belief Network model. The results of the research led to the conclusion
that a Bayesian Belief Network was a potentially useful tool for physician decision support process.

In addition, a Bayesian model for prediction of mental retardation in newborns (MENTOR) was developed from a large medical data set (Mani,
McDermott, and Valtorta, 1997). The system was developed using Hugin as a
tool and addressed the problem of missing values in big data sets arising from

incomplete clinical findings. The Bayesian Network structure and conditional probability values were modified under the guidance of domain experts, resulting in the research leading to successful implementation of Bayesian Network methodology using a domain expert's knowledge.

Bellazi and Riva (1995) described a methodology using a time series interpretation that used causal probabilistic models. In a case study, they applied this approach to a diabetes-monitoring domain. They found that a discrete-time/time-invariant dynamic system could easily be encoded within a Bayesian framework. More recently, Bellazi and Riva (1998) demonstrated a mechanism to learn Bayesian Network probabilities from longitudinal data and applied this technique to a medical domain problem.

Larizza et al. (1997) used a Temporal Abstraction mechanism to summarize a patient's behavior and explain the dynamics of evolution over a predefined time interval and interpreted the long-term monitoring data to explain the dynamics of the patient's evolution. They have suggested several procedures using Temporal Abstractions to include detection of interesting episodes, model data extraction, and time series analysis of abstracted events.

Furness, Kazi, Nicholson, Kirkpatrick, Taub, Davies, and Solez (1997) used a Bayesian Belief Network in grading ten different features in each of 100 transplant biopsies to determine whether rejection could be predicted. The data was used to develop a conditional probability matrix and was included in a computer program. The results indicated that Bayesian Belief Networks could be used for predicting early rejection in transplant pathology with improved accuracy.

The research of Chevrolat, Golmard, Ammar, Jouvent, and Boisvieux (1998) applied Bayesian networks-modeling to a multidimensional depression problem. The model depicts the characterization of the probabilistic model using expert knowledge to associate latent concentrations of neurotransmitters and symptoms. The Bayesian analysis is carried out using Gibb's sampling procedure. Results of this research can be enlarged to the central problem of managing latent variables in Bayesian Network modeling.

Miller and Seaman (1998) easily applied hierarchical Bayesian methods to a combination of medical doses problem, yielding the probability that a drug combination is superior to its components. Further, they present methods that may be implemented using readily available software for numerical integration, as well as ones that incorporate Markov Chain Monte Carlo methods.

Using a Bayesian Network to represent uncertainty and incomplete data for obesity, the study by Bunn, Du, Niu, Johnson, Poston, and Foreyt (1999) proved that a complex medical problem, such as predicting obesity, could be broken down into manageable sub-tasks and modeled.

A case study by Landrum and Normand (1999) presented Bayesian analysis used to develop medical guidelines based on expert opinion, using ordinal categorical data. They developed guidelines for the use of coronary angiography following an acute myocardial infarction (AMI) for 890 clinical indications using Bayesian models and statistical fit to investigate the appropriateness ratings obtained from a nine-member panel of experts.

Mezzetti and Robertson (1999) developed a Bayesian hierarchical model to estimate age-specific cancer incidence per year from age-specific cancer mortality. The Bayesian model includes the number of cases per year, considered as observations from a discrete-time stochastic process, following an autoregressive structure within a Poisson regression model. The model assumes that the survival probability among those with cancer is known. The methodology has also been investigated using lung cancer mortality data and parameter estimates obtained through Markov Chain Monte Carlo methods.

The number of full-scale belief network applications is very small compared to clinical decision support systems developed using other methods. This may be for several reasons: the theory of probabilistic networks is fairly new; methodologies for building probabilistic networks are lacking; the network models do not provide any means for exerting control over reasoning behavior; and there are no known special-purpose methods for evaluating network-based systems.

The Contribution This Study Will Make to the Field

It is evident from the above discussions that Bayesian inference is an appropriate tool for modeling the uncertainties involved in this research. This research reviewed, incorporated, and made an effort to improve upon these known models, which can be potentially used to assess medication compliance among patients receiving medical treatment for glaucoma.

As the model was found to be appropriate in predicting the noncompliance and the duration of elevated IOP among glaucoma patients with reasonable accuracy, it can be used with modifications and refinements for other diseases characterized by inherent uncertainties.

3

Methodology

In this chapter, the research methods employed in modeling the problem of noncompliance using a Bayesian Network are explained. Use of Expectation Maximization (EM) algorithm to predict the missing values with reasonable accuracy is also explained. The Hugin Lite system is used in a sample case of noncompliance to explicate its implementation. Further, the format for presenting the results, expected outcomes, and resources used in the research is explained. The chapter concludes with the methodologies to predict the reliability and verify the validity of the data obtained.

Research Methods to be Employed

In this research, Bayesian Networks were used to develop knowledge-based applications in glaucoma patient compliance that is characterized by inherent uncertainty. Increasingly, Bayesian Network techniques are being used to develop advanced knowledge-based Decision Support Systems to solve real world problems. Bayesian Networks are particularly useful for diagnostic applications and have been deployed in many systems.

Bayesian modeling can be used whenever classical knowledge-based systems (KBSs) might be used. Bayesian Networks (BNs) provide a more modular representation of uncertain knowledge making them easier to maintain and adapt to different contexts. Also, BNs provide more intuitive knowledge representation (cause-effect node diagrams) for domain experts, making it easier for them to be involved in maintaining a system.

Ideally, the primary purpose of a Bayesian model is to give probabilistic estimates for events that are not observable. The Bayesian model consists of three components: a qualitative model, probabilities, and an inference system. The model is an explicit representation of glaucoma patient compliance that can be developed using knowledge from the domain expert or from available data. These models can be validated by comparing them to the performance of the domain expert (Spiegelhalter, Dawid, Lauritzen, and Cowell, 1993).

Specific Procedures Employed

The following procedure was used in constructing a Decision Support System in this research:

1. Identify the various variables and their relationship to patient compliance

2. Develop a model in causal form

3. Use the Hugin system to construct a Bayesian Structure

4. Gather prior probabilities from the domain expert

5. Analyze the inference mechanism in the Bayesian model

6. Update the evidence based on clinical data, using a learning algorithm

7. Test and validate the model

Construction of the model began by acquiring a qualitative model from an ophthalmologist (domain expert) for patient compliance processes of glaucoma (the domain of interest). The qualitative model consists of variables that represent entities and directed arcs that connect the variables. The model was built in causal form, that is, connections in the network point from causes to effects.

Identifying the Variables of Interest

The events in BNs are organized into sets of variables; namely, information variables or hypothesis variables (Jensen, 1997). A hypothesis variable contains events that are not observable. An information variable contains events that provide the state of the hypothesis variable. After discussing with the domain expert, we identified seven variables in the model. Each variable, along with their states, is defined in the following section.

1. Non-Compliance (NComp)

   Non-Compliance is defined as the number of weeks a patient has not applied medication despite a physician's instructions. We categorized the duration of patient's compliance into four hypothesis events, namely: "*Weeks 1–6*", "*Weeks 7–12*", "*Weeks 13–18*", and "*Weeks 19–24.*"

2. Unhealthy Intraocular Pressure (UIOP)

   It is known that development of glaucoma is directly related to pressure inside the eye—the intraocular pressure (IOP). Pressure builds up in the eye when the clear liquid called the aqueous humor, which normally flows in and out of the eye, is prevented from draining properly. The resulting increase in pressure within the eye can damage the optic nerve cup. We defined UIOP as the number of weeks of unhealthy pressure in the patient's eye. We categorized the duration of patient's unhealthy IOP into four hypothesis events namely: "*Weeks 1–6*", "*Weeks 7–12*", "*Weeks 13–18*", and "*Weeks 19–24.*"

3. Optic Nerve Cup (DeltaCup)

   We defined the DeltaCup variable as the detrimental change in the size of the optic nerve cup between two consecutive visits due to the unhealthy pressure in the eye; the values are "*changed*" (yes) and "*not changed*" (no). The value "*changed*" (yes) indicates that the optic nerve cup is subjected to more damage. If the glaucoma is untreated, the optic nerve cup will become larger and deeper, indicating a worsening of the glaucoma. The optic nerve cup represents irreversible loss of optic nerve tissue.

4. Visual Field (DeltaVF)

   We defined the DeltaVF variable as the detrimental change in visual field readings from two consecutive visits due to the unhealthy pressure in the eye; the values are "*changed*" (yes) and "*not changed*" (no). The value "*changed*" (yes) indicates that the visual field is lost due to the unhealthy pressure. This test allows the physician to know if and how the patient's field of vision has been affected by glaucoma. The visual field reading is an important measure of the extent of damage to the optic nerve.

5.   Dilated Pupil (DP)

This examination provides a better view of the optic nerve to check for signs of damage in the patient's eye. The pupil is widened (dilated) by adding drops to the eye. The close-up vision may remain blurred for several hours after the examination. The visual field readings may be affected due to dilation. The values for this variable are *"Pupil dilated"* (yes) and *"Pupil not dilated"* (no).

6.   Cataract (CT)

A cataract is an eye condition where cloudiness, or opacity in the lens, blocks or changes the entry of light, affecting vision. This is one of the many conditions other than glaucoma that can cause poor vision. The patient's relative insensitivity to light due to glaucoma cannot be accurately determined in the presence of cataract(s); it is important to consider this factor. However, a cataract tends to produce a similar pattern of diffused visual field loss, while glaucoma tends to produce localized areas of visual field loss. The values for this variable are *"Patient has cataract"*(yes) and *"Patient doesn't have cataract"* (no).

7.   Learning Ability (LA)

Learning ability is a psychophysical ability of the patient to respond to the questions during the visual field test. The values for this variable are *"Patient has learning ability"* (yes) and *"Patient doesn't have learning ability"* (no).

Bayesian Network Structure

The dependencies among the variables are identified as causal arrows pointing from causes to effects. The direction of the arc indicates that there is some form of influence or dependency between the variables. Figure 1 represents a Bayesian Network of the seven variables interacting with each other in the domain of our interest. Learning ability has a direct influence on the patient's compliance and also on visual field reading. Other variables that can affect visual field reading are Cataract and Dilated Pupil. When the patient has not

complied with the medication, unhealthy intraocular pressure builds up causing the visual field loss and optic nerve cup damage. This model is a directed
acyclic graph (DAG), which means there is no cyclic path from any node.
Although the network is not singly connected, containing only seven variables,
it may allow us to compute exact posterior probabilities in polynomial time.

For this model, we needed prior probabilities for three variables: Cataract,
Learning Ability, and Dilated Pupil. Conditional probabilities were then
sought for Noncompliance, given Learning Ability for two events—namely
"*with ability*" and "*without ability*,"—and the conditional probabilities for
Unhealthy IOP, given Noncompliance. Finally, we assessed the conditional
probability of Delta VF, given an Unhealthy IOP, Learning Ability, Cataract,
and Dilated Pupil.

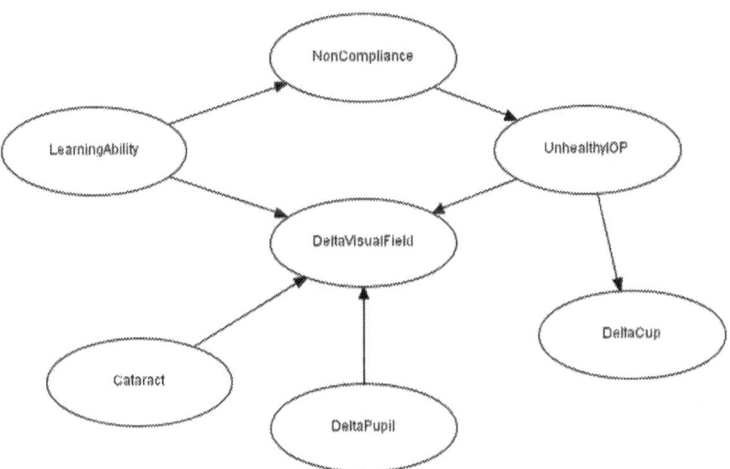

Figure 1. Bayesian model showing medication compliance in glaucoma
domain.

Prior Probabilities or Subjective Estimates

We needed the following probabilities for Figure 1:

1. Learning Ability Node:

    P(Learning ability): Prior probability of the patient's ability to learn.

2. Cataract Node:

    P(Cataract): Prior probability of patient having a cataract.

3. Dilated Pupil Node:

    P(Dilated pupil): Prior probability of patient having pupil dilated.

4. Non-Compliance Node:

    •   P(Non-Compliance | Learning Ability = yes): Prior probability of
        Non-Compliance given that the patient has the Learning Ability.

    •   P(Non-Compliance | Learning Ability = no): Prior probability of
        Non-Compliance given that patient does not have the learning ability.

5. Unhealthy IOP Node:

    P(Unhealthy IOP | Non-Compliance): Prior probability of patient hav-
    ing unhealthy IOP given Non-Compliance for different events.

6. Delta Cup Node:

    P(DeltaCup | Unhealthy IOP): Prior probability of changes in Delta Cup
    given Unhealthy IOP.

7. Visual Field Node:

    This node has multiple parents and a full joint probability is required over
    all states of its parents.

Tables 1 through 10 are the prior probabilities that were provided by the
ophthalmologist's (domain expert) records. In the Bayesian model, a probability describes the strength of the belief (prior probabilities), which a domain
expert can justifiably hold that a certain statement of fact is indeed true. These
probabilities can be derived from patient databases if such data set is available.

Table 1. Prior P(Learning ability)

Events	Probabilities
Learning ability = "Yes"	0.85
Learning ability = "No"	0.15

Table 2. Prior P(Non-Compliance| Learning ability) or P(Ncomp| Learning ability)

Events	Probabilities	
Duration in weeks	Learning ability = "Yes"	Learning ability = "No"
NComp = "1 to 6"	0.5	0.03
NComp = "7 to 12"	0.25	0.07
NComp = "13 to 18"	0.15	0.3
NComp = "19 to 24"	0.1	0.6

Table 3. Prior P(Unhealthy IOP| Ncomp) or P(UIOP|NComp)

Events	Probabilities			
In weeks	Ncomp = "1 to 6"	NComp = "7 to 12"	NComp = "13 to 18"	NComp = "19 to 24"
UIOP = "1 to 6"	0.9	0.05	0.03	0.02
UIOP = "7 to 12"	0.05	0.9	0.02	0.03
UIOP = "13 to 18"	0.03	0.02	0.9	0.05
UIOP = "19 to 24"	0.02	0.03	0.05	0.9

Table 4. Prior P(DeltaCUP|UIOP)

Events	Probabilities			
In Weeks	UIOP = "1 to 6"	UIOP = "7 to 12"	UIOP = "13 to 18"	UIOP = "19 to 24"
DeltaCup = "Yes"	0.2	0.5	0.7	0.95
DeltaCup = "No"	0.8	0.5	0.3	0.05

Table 5. Prior P(Dilated Pupil)

Events	Probabilities
Dilated Pupil = "Yes"	0.4
Dilated Pupil = "No"	0.6

Table 6. Prior P(Cataract)

Events	Probabilities
Cataract = "Yes"	0.6
Cataract = "No"	0.4

Table 7. Prior P(DeltaVF|UIOP, Learning ability, Dilated Pupil = "Yes", Cataract = "Yes")

Events		Probabilities			
	In weeks	UIOP = "1 to 6"	UIOP = "7 to 12"	UIOP = "13 to 18"	UIOP = "19 to 24"
DeltaVF = "Y"	Learning ability = "Yes"	0.45	0.6	0.85	0.95
DeltaVF = "N"		0.55	0.4	0.15	0.05
DeltaVF = "Y"	Learning ability = "No"	0.5	0.65	0.9	0.95
DeltaVF = "N"		0.5	0.35	0.1	0.05

Table 8. Prior P(DeltaVF|UIOP, Learning ability, Dilated Pupil = "Yes", Cataract = "No")

Events		Probabilities			
	In weeks	UIOP = "1 to 6"	UIOP = "7 to 12"	UIOP = "13 to 18"	UIOP = "19 to 24"
DeltaVF = "Y"	Learning ability = "Yes"	0.35	0.5	0.75	0.95
DeltaVF = "N"		0.65	0.5	0.25	0.05
DeltaVF = "Y"	Learning ability = "No"	0.4	0.55	0.8	0.95
DeltaVF = "N"		0.6	0.45	0.2	0.05

Table 9. Prior P(DeltaVF|UIOP, Learning ability, Dilated Pupil = "No", Cataract = "Yes")

Events		Probabilities			
	In weeks	UIOP = "1 to 6"	UIOP = "7 to 12"	UIOP = "13 to 18"	UIOP = "19 to 24"
DeltaVF = "Y"	Learning ability = "Yes"	0.25	0.4	0.7	0.9
DeltaVF = "N"		0.75	0.6	0.3	0.1
DeltaVF = "Y"	Learning ability = "No"	0.3	0.45	0.75	0.95
DeltaVF = "N"		0.7	0.55	0.25	0.05

Table 10. Prior P(DeltaVF|UIOP, Learning ability, Dilated Pupil = "No", Cataract = "No")

Events		Probabilities			
	In weeks	UIOP = "1 to 6"	UIOP = "7 to 12"	UIOP = "13 to 18"	UIOP = "19 to 24"
DeltaVF = "Y"	Learning ability = Yes"	0.4	0.55	0.8	0.95
DeltaVF = "N"		0.6	0.45	0.2	0.05
DeltaVF = "Y"	Learning ability = No"	0.2	0.35	0.65	0.85
DeltaVF = "N"		0.8	0.65	0.35	0.15

Inference System in a Bayesian Environment

Having specified the Bayesian model through a domain expert's opinion, the HUGIN model's inference techniques were used to propagate probabilities through the model (Jensen, 1997). Given the evidence for observed events, the probabilities of other events that have not been observed were propagated. For example, in Figure 1 we entered evidence that there was a change in both the DeltaCup and DeltaVF. Using the Bayes theorem, it was then possible to update the values of all of the other probabilities in the BN.

Figure 2 shows the initial state of the BN with no evidence entered. Each node in the network has a histogram showing the probability of node states being true. The prior belief led us to believe that the patient "noncomplied" for one to six weeks and that the unhealthy pressure built up during that period. This built-in bias leaned toward noncompliance, even though we knew nothing about the patient.

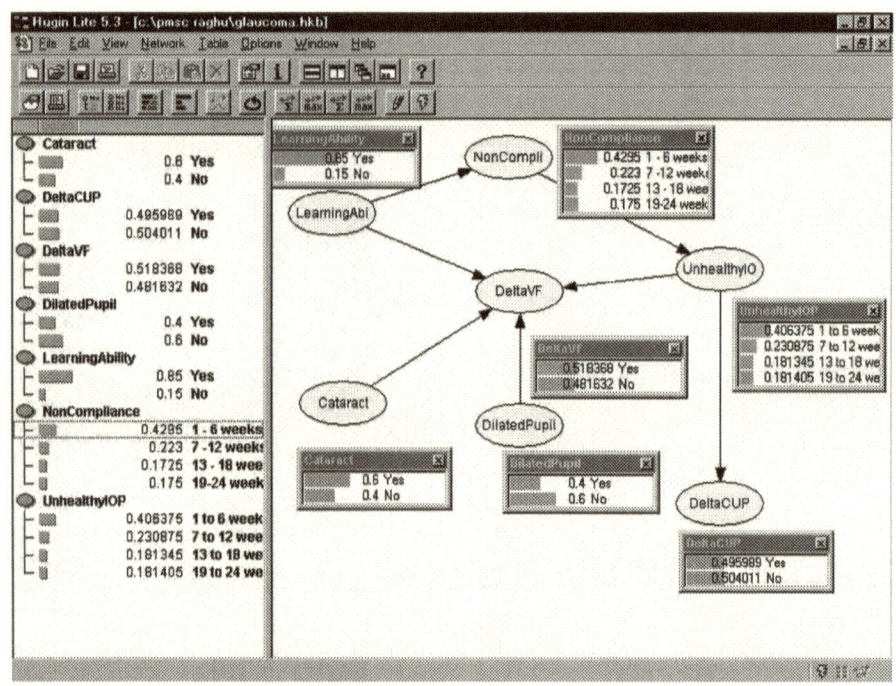

Figure 2. Bayesian Model showing medication compliance—Initial state.

In Figure 3, we entered evidence that there was no change in either DeltaVF or DeltaCup in a patient who had complied with the physician instructions. When the evidence was entered, the probability became 100% for that variable and the histogram bar turned from green to red. Note how this led to significant changes in the Unhealthy IOP variable. The probability of the unhealthy pressure for this patient was almost zero.

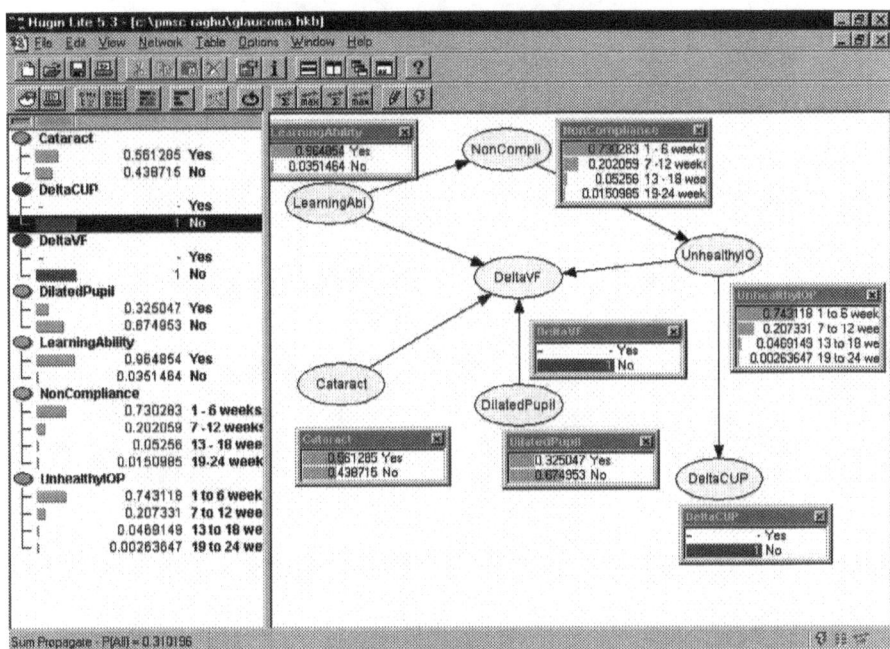

Figure 3. Bayesian Model showing medication compliance—"Complied" Patient

Figure 4 shows that we have evidence of DeltaVF and DeltaCup, along with Learning Ability. Finally, we saw dramatic increases in a belief of Noncompliance and Unhealthy IOP. The evidence for DeltaVF also propagated back to the process nodes. Our belief in a high Unhealthy IOP was drastically increased; this in turn led to a strong belief in the Noncompliance of the patient.

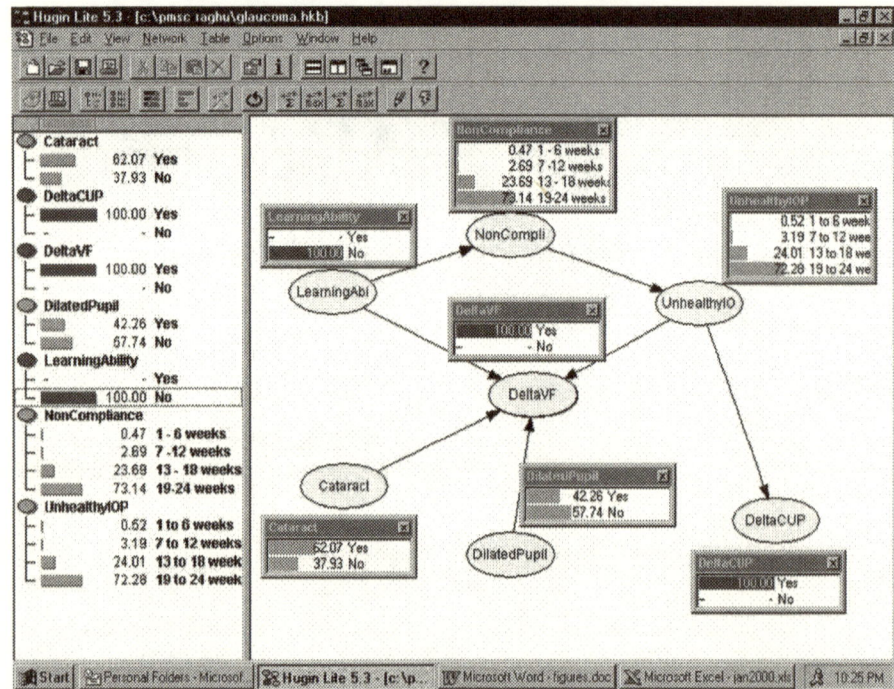

Figure 4. Bayesian Model showing medication compliance—"non-complied" Patient

The major benefit of a Bayesian inference system is that it explicitly describes the fact that observation alone cannot predict the probability of unobserved events, without some prior information. A significant problem for BNs is the difficulty of performing the propagation, especially when the network model is multi-connected (there is more than one undirected path between some nodes). In fact, the general problem of performing such computations is known to be NP-hard (Cooper and Herskovits, 1992). The Hugin tool, which provided a graphical interface for the Bayesian propagation, made it feasible to model uncertain inference problems. Building a BN required a large number of probability values, and certain assumptions were made to reduce the number of assessments by assuming conditional independence.

Deriving Posterior Probabilities from Data.

The glaucoma patient compliance problem has inherent uncertainty and Bayesian Belief Networks seem to be an appropriate tool for this problem. Once the model was constructed using the Hugin system and prior probabilities were specified, the next step was to update the priors using clinical data. As in many real world situations, the glaucoma patient compliance model had two variables (Noncompliance and Unhealthy IOP) that did not have any data. The data for the duration of noncompliance, and the duration of elevated IOP in the six months preceding the clinical visit were not available to the doctor for making useful inferences while examining the patient. The missing data complicated the problem of deriving posterior probabilities. Missing or incomplete data poses a serious problem: the more data that are missing, the more pressing the need to address the problem of an incomplete data set.

Few studies have addressed the problem of missing data in general, and Bayesian Belief Networks, in particular. It is essential to account for the missing data to develop a useful model. In this study we were aware that a model without proper accounting of missing data would produce biased and unreliable results. But during the last 30 years, techniques have been developed and refined to solve the problem of missing data.

The principal methods developed for accounting for missing data are: the Expectation-Maximization (EM), by Dempster, Laird and Rubin in 1977; Gibbs Sampling, by Geman and Geman (1984); and the Multiple Imputation, by Rubin (1987). Some further work has been done in the field by adapting the EM and Gibbs Sampling to solve Bayesian problems. The researchers assumed that the structure was known and did not fully explore the task of learning the structure or the parameters from incomplete data (Singh, 1997). Current research includes learning parameters from incomplete data using gradient methods (Binder, Koller, Russell, and Kanazawa, 1997; Thiesson, 1995) and learning Bayesian structure (Friedman, 1997; Friedman, 1998; Meila and Jordan, 1998; Thiesson, Meek, Chickering, and Heckerman, 1998).

Singh (1997) proposed an iterative algorithm that successively refined the learned Bayesian Network by combining Expectation-Maximization and Imputation techniques. Experimental results showed that the quality of the distribution was closer to the real values than the ad-hoc procedures commonly used for missing data. The algorithm proposed by Singh (1997) was used in this study.

The most appropriate way to handle missing or incomplete data depends upon how data set points became missing. Little and Rubin (1987) define three unique types of missing data mechanisms:

1. Missing Completely at Random (MCAR)

2. Missing at Random (MAR)

3. Not Missing at Random (NMAR)

Missing Completely At Random (MCAR) means that the data set is unrelated to the values of the variables (Little and Schenker, 1995). In this research, we explored relationships between glaucoma noncompliance variables such as, Learning Ability, Visual Field, Cataract, Noncompliance and Unhealthy IOP. The missing data pattern was MCAR because Noncompliance and Unhealthy IOP were unrelated to the values of Cataract, Learning Ability or Visual Field. Thus, values for Noncompliance and Unhealthy IOP were always missing for this dataset.

Missing At Random (MAR) means that the distribution, given the observed variables, depends only upon variables that are recorded in the data set (Little and Schenker, 1995). Continuing with the noncompliance example in the previous paragraph, consider that Visual Field was fully observed, and that a value for Cataract was sometimes missing, was a MAR problem.

Not Missing at Random (NMAR) means that the missing data is dependent on the value of the missing component itself, and possibly on the value of observed data. The type of missing data mechanism plays a vital role in determining the type of problems that can addressed by various learning algorithms (Singh, 1997).

The above categories can be explained using a matrix representation. Let

$$D = \langle D^{obs}, D^{mis} \rangle = (d_{li})_{m \times n}$$

D^{obs} and D^{mis} represent the observed and missing component of the data D, and d_{li} represent the value of ith attribute in l^{th} case. Let θ represent the parameters governing the generation of the data, and $P(D|\theta) \equiv P(D^{obs}, D^{mis}|\theta)$ denote the density of joint distribution for D^{obs} and D^{mis}.

The missing data mechanism can be represented as a matrix $R = (r_{li})_{mxn}$ where r_{li} is 1 if d_{li} is observed or 0, if it is otherwise. If a set of parameters ψ governs the distribution for the missing data mechanism, the equation combining R and D can be rewritten as Equation 1:

$$P(D^{obs}, D^{mis}, R \mid \theta, \psi) = P(D^{obs}, D^{mis} \mid \theta)P(R \mid D^{obs}, D^{mis}, \psi) \ldots\ldots\ldots(1)$$

If the probability of the missing data is independent of the observed D^{obs} and missing data D^{mis}, the missing data is said to be MCAR. Mathematically, it can be expressed as:

$$P(R \mid D^{obs}, D^{mis}, \psi) = P(R \mid \psi)$$

If the probability that missing data is independent of the missing components and may depend on the values of the observed variables, it is said to be MAR. It can be expressed mathematically as:

$$P(R \mid D^{obs}, D^{mis}, \psi) = P(R \mid D^{obs}, \psi)$$

If the probability that missing data is dependent on the missing component and probably some of the observed variables, it is said to be NMAR.

The likelihood L of θ being dependent on the observed components of the data D^{obs}, considering the missing data can be mathematically expressed as:

$$L(\theta, \psi \mid D^{obs}, R) \propto P(D^{obs}, R \mid \theta, \psi).$$

In our case, the missing values, the period of noncompliance and the duration of elevated IOP were missing parameters. The missing component of noncompliance with medication was not dependent on any of the missing parameters, namely the noncompliance itself and the elevated IOP. The missing component of noncompliance was not a function of the observed parameters, which are the changes in the visual field and the optic nerve cup. The same holds true for the missing component of the elevated IOP in the eye for an unknown duration in the six months preceding the clinical visit. Hence, it was evident that the procedure to solve MCAR routines should be used for solving the problem of interest.

The EM algorithm is a general iterative algorithm for maximum likelihood estimation in problems of missing data. The algorithm estimates the likelihood by alternating between the Expectation Step (E) and Maximization Step (M). The EM methods were used over the Gibbs Sampling because of the tendency of the EM routines to converge quickly, compared to the Gibbs. However, EM algorithms have the disadvantage of finding only local maxima (Singh, 1997).

Lauritzen (1995) proposed a likelihood function given by:

$$L_B(\theta \mid D) \propto P_B(D \mid \theta) = \prod_{i=1}^{n} \prod_{j=1}^{q_i} \prod_{k=1}^{r_i} \theta_{ijk}^{N_{ijk}}$$

where B is the Bayesian Network Structure, n is the number of attributes, q_i is the number of possible instantiations of the parents, $pa(i)$, of the ith attribute, r_i is the number of attribute i, N_{ijk} is the number of cases in which the attribute i is instantiated to k, while $pa(i)$ is instantiated to j, and $\theta = (\theta ijk)_{n \times q(i)}$ $\times r(i)$ are the conditional probabilities.

The **E** step determines the conditional expectations of the complete-data likelihood, given the current values of the parameters and the observed components of the data. The previous statement can be mathematically written as:

$$E(N_{ijk} \mid D^{obs}, \theta) = \sum_{l=1}^{m} E_\theta(\chi_{ijk}^{l} \mid d_l^{obs})$$

where d_l^{obs} is the observed component of the lth case, and χ_{ijk}^{l} can be defined as:

$$E_\theta(\chi_{ijk}^{l} \mid d_l^{obs}) = \begin{cases} 1, & i, pa(i) & observed; i = k \ \& \ pa(i) = j \\ 0, & i, pa(i) & observed; i \neq k \cdot or \cdot pa(i) \neq j \\ P_B \ (i = k, pa(i) = j \mid d_l^{obs}, \theta) \end{cases}$$

To determine the value of P_B $(i = k, pa(i) = j \mid d^{obs}_l, \theta)$, the Bayesian Network is instantiated with d^{obs}_l and the inference is carried out. The **M** step consists of using these estimated values to compute the different θ*ijk* values. It can be expressed as:

$$\theta_{ijk} = \frac{E_\theta(N_{ijk} \mid D^{obs}, \theta)}{E_\theta(N_{ij} \mid D^{obs}, \theta)} \; whereN_{ij} = \sum_k N_{ijk}$$

The EM Algorithm for Bayesian Networks (Singh, 1997).

An algorithm using a combination of EM and imputation techniques to refine the variables was used. The current estimates of the structure and its incomplete data were used as a starting point to refine the conditional probabilities. Then new values were imputed for missing data points by sampling from the new estimate of the conditional probabilities. The network was refined with the help of standard algorithms for learning Bayesian Networks from complete data. Since these data are not fill-in values, but rather estimates of the conditional probabilities, they cannot be directly used. However, it can be used for refining the structure of the model and the conditional probabilities.

When convergence was observed, the values obtained through use of EM algorithms were found to be closer to real world values than the ad-hoc assumption of the missing values or completing the network using ad-hoc assumptions. The algorithm suggested by Singh (1997) was used in this research to complete the network and populate the unknown variables (missing values) with more realistic estimates than the ad-hoc assumptions made by the ophthalmologists about the patient's condition.

Format for Presenting the Results

The contribution of this research will be presented via details about the proposed model through enumerating: the variables in the patient compliance process, a description of the inference mechanism, a statement of the implications of the results, and a summary of this research and further research suggestions.

Projected Outcomes

An extensive search revealed that no valid tools were available to predict non-compliance among the glaucoma patients, or the duration of their elevated IOP, with reasonable accuracy. The model developed in this research was a first step (albeit modest) in determining these factors with acceptable accuracy.

The model ideally provided approximate estimates of the duration of non-compliance and elevated IOP. This was beneficial in understanding the rate at which the disease worsens due to either noncompliance of medication or duration of elevated IOP, or both.

Resource Requirements

The resources that were required to complete this research involved an educational version of the Hugin (available at www.hugin.dk) software. Higher level programming (Visual C++) was involved in the process. The University of South Carolina's Artificial Intelligence Laboratory has the Hugin system installed on its workstations. Permission was obtained to use the system. Patient data was available from the University of South Carolina's Eye institute. Permission was obtained from the Internal Review Board (IRB) to use their data for research purposes.

Reliability and Validity

Reliability in this research refers to the consistency of the results provided by the Bayesian framework. The model output is expected not to vary from the domain expert's evaluation and assessment of patient compliance. Whether a patient is complying with physician's instructions in glaucoma domain involves uncertainty. Further, the patient data may be missing or incomplete. The Bayesian model, in its theoretical construct and practical application, ideally lends itself to evaluate such problems. The Bayesian Belief Network was developed based on the understanding of the domain expert. The cause and effect relations are not completely deterministic as in rule-based Expert Systems. The strengths of these relations were modeled with probabilities. The BN uses classical probability theory and models the dependencies from the structural description of the problem.

An incomplete dataset was prior knowledge in this research. The probability of the missing data was independent of the data observed and categorized as a missing at random (MCAR) problem. The Expectation Maximization (EM) algorithm was the preferred approach to solve the problem of missing data (Lauritzen, 1995; Singh, 1997). Singh (1997) addressed the problem of missing data for Bayesian Network using EM algorithm. Model reliability was enhanced via the use of the EM algorithm. Hence, a problem that can be modeled as a cause and effect relationship with missing data can be reliably modeled using an EM algorithm adapted for solving Bayesian Belief Network problems.

A large number of hospitals treat glaucoma patients but many are reluctant, or even refuse, to share the information, citing privacy and doctor-patient privileges. Hence, the amount of data made available for this research by the hospitals was limited in the number of patient cases. Since this research is a first step in modeling noncompliance among glaucoma patients, nearly three fourths (75%) of the patients' data was used for refining the model and understanding the relationships and the network (the remaining were used to validate the model). The ophthalmologist was asked to comment on patient noncompliance and the approximate duration of elevated IOP. The ophthalmologist's subjective evaluation was compared to the predicted values of the model. This procedure helped to verify the appropriateness of the model and identified ways to refine it.

Validity in this research refers to the appropriateness, meaningfulness, and usefulness of the model output. The present research relied upon "criterion validity" which compares "the degree to which information provided by an instrument agrees with information obtained from other sources" (Fraenkel and Wallen, 1996). The Bayesian model was used as the "criterion instrument." To assess research validity, the results obtained through the model were compared against the domain expert's evaluation of each patient case. Since the results provided by the instrument agreed with the information provided by the domain expert, the Bayesian framework for the glaucoma model is considered reasonably valid.

Summary

In this chapter, the methodologies to solve the problem of noncompliance among glaucoma patients using Bayesian Belief Networks modified with EM

algorithm were explained. The projected data outcome, reliability of the
methodology and ways to verify validity of the results, were also discussed.

4

Results

In this chapter, the results obtained from the analysis using the Hugin Lite Bayesian Belief Network tool is presented in the form of posterior probabilities of the known variables and the determined probabilities for the unknown variables. Further, the data is analyzed to derive meaningful medical deductions. These deductions, insights into the problem of noncompliance among glaucoma patients, and the factors affecting them are also discussed.

Derivation of Posterior Probabilities

To refine the variables, we used an implementation developed by Singh (1998) that uses an algorithm that combines EM and imputation techniques. The prior estimates of the Bayesian Structure and its incomplete data were used as a starting point to refine the conditional probabilities. Then new values were imputed for missing data points by sampling from the new estimate of the conditional probabilities. The network was refined with the help of standard algorithms for learning Bayesian Networks from complete data. The EM program was written in C programming language (ANSI standard) and modified to run in the Microsoft Visual C++ 6.0 environment.

The EM Implementation

To implement the EM program, the Hugin inference engine, offered by the Hugin Application Program Interface (API), (Hugin, 2000) was used. The

Hugin API is provided in the form of a library that can be linked into applications written using the C++ programming language; therefore, it can be included in the implementation. The Hugin API contains a high performance inference engine that can be used as the core of decision support systems using a Bayesian framework. The EM program mainly consists of three programs: an interface program, the main program, and a utilities program. The interface program collects the input for the main program, and using function calls, passes the input to the main program. The utility program computes the expected clique marginal counts and the log-likelihood estimates. The EM program requires two input files: a quantitative conditional probability network (CPN) file describing the BN structure with prior probabilities, and a dataset containing patient cases with observed and missing data. The program produces two output files: a CPN file containing the BN structure with posterior probabilities, and a data file showing the number of iterations before the convergence. Figure 5 shows the process of the EM implementation.

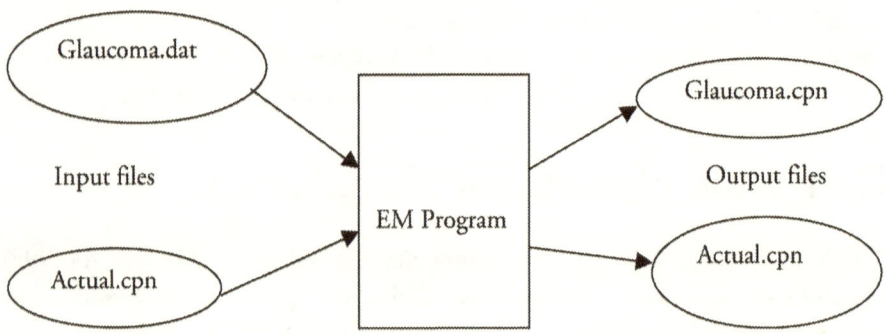

Figure 5. EM implementation with input and output files.

The procedure for creating the input files is mentioned below. The model was developed using the Hugin Lite based on the dependent variables and their prior probabilities from the domain expert. The Hugin system allowed us to save the structure and probabilities in a text file (actual.net or actual.cpn) that can be used as an input file for the EM program and the Hugin API. The following section describes the input files.

Input Domain NET File (actual.cpn)

For the EM program to run, it was necessary to specify the qualitative part of the BN with prior probabilities. This file consists of variables in the model along their states.

The format of the states had to be in integer values; the states "*Yes*" and "*No*" were renamed to "1" and "2" respectively. The values for variables such as Noncompliance were changed to "1," "2," "3," and "4" for periods of "*1 to 6 weeks*," "*7 to 12 weeks*," "*13 to 18 weeks*," and "*19 to 24 weeks*," respectively. Further, for programming purposes, the variables (nodes) were changed from their actual names to N1, N2, N3 to N7, while keeping the order of the nodes. Table 11 indicates the new names of the variables used in the EM program.

Table 11. Mapping of variable names used in EM program.

Node Numbers in actual.cpn file	Actual Variable Names in the Bayesian Model
N1	Learning Ability
N2	Cataract
N3	Dilated Pupil
N4	Non-Compliance
N5	Unhealthy IOP
N6	DeltaCUP
N7	DeltaVF

The input file "actual.cpn" is shown in Appendix A. This file describes the BN structure, nodes, states, and parents for each node along with the conditional probabilities. For details on CPN files and NET language, consult the reference manual for the Hugin API (Hugin, 2000).

Input Datafile (glaucoma.dat)

Another input file for the EM program, galucoma.dat, contains the observations. This datafile must have the following format: The first line consists of the number variables in the Bayesian model followed by the number of states each variable has. Variables have to be in the same order as in the other input (actual.cpn) file. Each value has to be separated by a space. The next line is the number of cases in the data set. In the following lines, the observations are entered. Values for these observations have to follow the same format as in the cpn file. The values of observed data are separated by a space, and a missing value is denoted by "0". The datafile (glaucoma.dat) containing the patient data for glaucoma noncompliance process is illustrated in Appendix A.

Once the program is run, the output file containing posterior probability distribution is generated and can be used in the Hugin system. Another output file called glaucoma.out was created to show the iterations of the EM before the convergence. The following section describes the two output files.

Final Probability Distribution File (glaucoma.cpn)

This file is in the same format as the input cpn file and contains the BN structure and final distributions that can be used directly in Hugin. The names of the variables were re-mapped to their original descriptive values, along with their states and coordinates of the nodes in the model. Appendix B displays the glaucoma.cpn file containing the posterior probabilities. In addition, the Final.net file shows the variables re-mapped to their original descriptive values. This file was used in Hugin for further testing.

Output File Showing Iterations (glaucoma.out)

The output file "glaucoma.out" illustrates the number of each iteration and their corresponding log-likelihood value. It shows the largest final log-likelihood value where the iteration converged.

The EM program has been developed for the purpose of parameter estimation within the Bayesian models on the basis of possible incomplete or missing data. This implementation could be an automatic tool for acquiring the quantitative part of a Bayesian Network for use in the Hugin system for data sets containing incomplete or missing data. This program can handle all combinations of missing or incomplete data. The current implementation of this EM program was run in the Microsoft Visual C++ 6.0 environment on a Windows platform.

Data Analysis

In this section, Tables 12 to 21 contain the posterior probabilities generated from the Hugin Lite Bayesian tool. In this study, a preliminary data set of 15 patients was used to refine the model in the Hugin Lite tool. This data helped us arrive at the posterior probabilities shown in tables 12 to 21. As a means of verifying the data, an additional set of four patients was entered into the system and the domain expert verified the results.

Posterior Probability Tables

Tables 12 through 21 are the posterior probabilities that were derived from the available patient dataset. In the Bayesian model, a probability describes the strength of the relationship between variables.

Table 12. Posterior P(Learning ability)

Events	Posterior Probabilities
Learning ability = "Yes"	0.72
Learning ability = "No"	0.28

Table 13. Posterior P(Non-Compliance| Learning ability) or P(Ncomp| Learning ability)

Events	Posterior Probabilities	
Duration in weeks	Learning ability = "Yes"	Learning ability = "No"
NComp = "1 to 6"	0.25	0.25
NComp = "7 to 12"	0.25	0.25
NComp = "13 to 18"	0.25	0.25
NComp = "19 to 24"	0.25	0.25

Table 14. Posterior P(Unhealthy IOP| Ncomp) or P(UIOP|NComp)

Events	Posterior Probabilities			
In weeks	Ncomp = "1 to 6"	Ncomp = "7 to 12"	NComp = "13 to 18"	NComp = "19 to 24"
UIOP = "1 to 6"	0.24	0.24	0.24	0.24
UIOP = "7 to 12"	0.25	0.25	0.25	0.25
UIOP = "13 to 18"	0.25	0.25	0.25	0.25
UIOP = "19 to 24"	0.26	0.26	0.26	0.26

Table 15. Posterior P(DeltaCUP|UIOP)

Events	Posterior Probabilities			
In Weeks	UIOP = "1 to 6"	UIOP = "7 to 12"	UIOP = "13 to 18"	UIOP = "19 to 24"
DeltaCup = "Yes"	0.64	0.66	0.67	0.69
DeltaCup = "No"	0.36	0.34	0.33	0.31

Table 16. Posterior P(Dilated Pupil)

Events	Posterior Probabilities
Dilated Pupil = "Yes"	0.39
Dilated Pupil = "No"	0.61

Table 17. Posterior P(Cataract)

Events	Posterior Probabilities
Cataract = "Yes"	0.5
Cataract = "No"	0.5

Table 18. Posterior P(DeltaVF|UIOP, Learning ability, Dilated Pupil = "Yes", Cataract = "Yes")

Events		Posterior Probabilities			
	In weeks	UIOP = "1 to 6"	UIOP = "7 to 12"	UIOP = "13 to 18"	UIOP = "19 to 24"
DeltaVF = "Y"	Learning ability = "Yes"	0.42	0.42	0.42	0.41
DeltaVF = "N"		0.58	0.58	0.58	0.59
DeltaVF = "Y"	Learning ability = "No"	0.63	0.63	0.64	0.65
DeltaVF = "N"		0.37	0.37	0.36	0.35

Table 19. Posterior P(DeltaVF|UIOP, Learning ability, Dilated Pupil = "Yes", Cataract = "No")

Events		Posterior Probabilities			
	In weeks	UIOP = "1 to 6"	UIOP = "7 to 12"	UIOP = "13 to 18"	UIOP = "19 to 24"
DeltaVF = "Y"	Learning ability = "Yes"	0.55	0.55	0.56	0.56
DeltaVF = "N"		0.45	0.45	0.44	0.44
DeltaVF = "Y"	Learning ability = "No"	0.65	0.66	0.67	0.68
DeltaVF = "N"		0.35	0.34	0.33	0.32

Table 20. Posterior P(DeltaVF|UIOP, Learning ability, Dilated Pupil = "No", Cataract = "Yes")

Events		Posterior Probabilities			
	In weeks	UIOP = "1 to 6"	UIOP = "7 to 12"	UIOP = "13 to 18"	UIOP = "19 to 24"
DeltaVF = "Y"	Learning ability = "Yes"	0.55	0.55	0.56	0.56
DeltaVF = "N"		0.45	0.45	0.44	0.56
DeltaVF = "Y"	Learning ability = "No"	0.5	0.5	0.5	0.5
DeltaVF = "N"		0.5	0.5	0.5	0.5

Table 21. Posterior P(DeltaVF|UIOP, Learning ability, Dilated Pupil = "No", Cataract = "No")

Events		Posterior Probabilities			
	In weeks	UIOP = "1 to 6"	UIOP = "7 to 12"	UIOP = "13 to 18"	UIOP = "19 to 24"
DeltaVF = "Y"	Learning ability = Yes"	0.5	0.5	0.5	0.5
DeltaVF = "N"		0.5	0.5	0.5	0.5
DeltaVF = "Y"	Learning ability = No"	0.53	0.54	0.55	0.56
DeltaVF = "N"		0.47	0.46	0.45	0.44

Model Evaluation

In evaluating the Bayesian framework, it is important to compare the behavior of the model (glaucoma model) to the real-life system that is being modeled (strategies used by ophthalmologists) and adjust the model according to the evaluation. Because of the inherent uncertainty in the glaucoma noncompliance process, two equal prior states of the system may result in different observed results due to the stochastic influence. However, this model was developed to include uncertainty in noncompliance behavior. The results show a probability distribution for two sets of observed results: one from the model and another from the ophthalmologist (domain expert). In this research, the Bayesian model was evaluated by comparing the true observations to the predicted probability distributions.

Verification with Additional Set of Data

The model was developed using a limited dataset and verified with four sets of new data. A great amount of data was not available due to privacy regulations and patient-doctor privileges. If larger datasets were used in the model, the probabilities could be further refined for increased predictability.

These values were then shown to the domain expert for an opinion. The domain expert felt that the predictions of the model correlated with the understanding of domain experts and was in line with the beliefs of the medical paradigm. The domain expert felt that the model could be further refined with a larger dataset to predict with greater accuracy. Comparisons between the domain expert's expected values and the predicted values of the model are presented in Table 22. A detailed analysis of each of the cases is presented afterward.

Table 22. Comparison of predicted and expected values of 4 test cases.

Case #	Learning Ability	Dilated Pupil	Cataract	Delta Cup	Delta VF	Ncomp (Prediction of the model)	Domain Experts Opinion
1	Yes	No	No	Yes	No	1 to 6 weeks	Agreed
2	Yes	No	Yes	No	Yes	1 to 6 weeks	Agreed
3	Yes	No	Yes	Yes	Yes	1 to 6 weeks	Agreed
4	Yes	No	Yes	Yes	Yes	1 to 6 weeks	Disagreed

A total of four patient cases were run on the Bayesian model to arrive at estimates of noncompliance. In the first case, the patient has the learning ability (value = "yes") with no cataract (value = "no") and dilated pupil (value = "no"), but a change in cup size (value of DeltaCup = "yes") and no change in visual field reading (value of DeltaVF = "no"). For this patient, the probability of having not complied for 1 to 6 weeks is high. The ophthalmologist was then asked to agree or disagree with the result. The expert agreed with the model's assessment. The expert was then asked to comment on the remaining three cases. The expert was in agreement with three out of the four cases, but t was not in agreement with Case #4. Although Case #4 is similar to Case #3 including the results, but the expert disagreed on the results of the Case #4. After reviewing the medical chart of Case #4, it was found that the variance of visual field and cup size was high compared to Case #3. In the Bayesian model, the designated values for visual field and cup size are *"changed"* or *"not changed."* In order to simplify the complexity of the model, the scope of the research was limited to the values of *"changed"* or *"not changed."* But the values for these two variables can be further refined to have ranges of variance, such as *"10% changed,"* *"20% changed,"* and so on. The expert agreed with the enhancement. This is covered in the findings as a way to enhance the present research.

Figure 6 shows the state of the learned Bayesian Network with evidence from Case #1 entered. The prediction from the model is that the patient was noncompliant for one to six weeks.

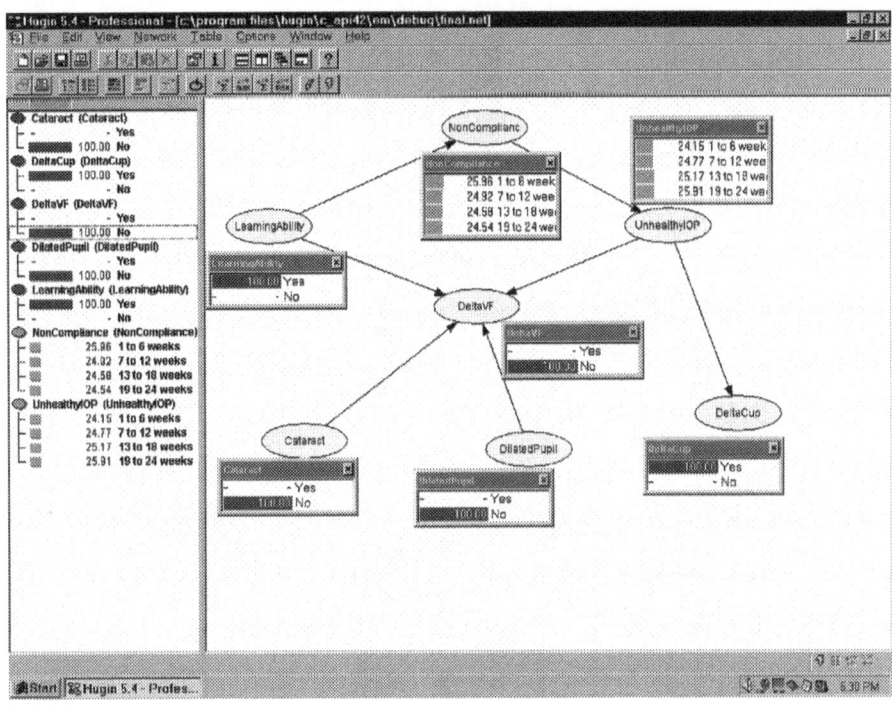

Figure 6. Bayesian model showing the evidence entered from case #1

Figure 7 shows the state of the Bayesian Network with the evidence from case #2 entered. The prediction from the model is that the patient was "noncompliant" for one to six weeks and an unhealthy pressure built up during that period.

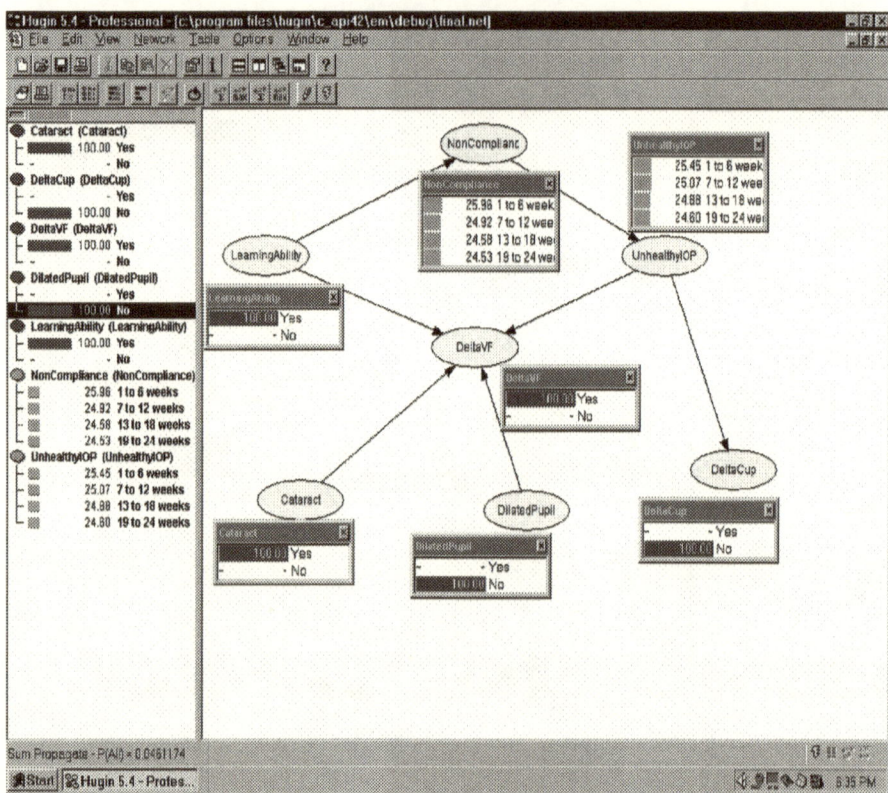

Figure 7. Bayesian Model Showing evidence entered from Case #2

Figure 8 shows the state of the Bayesian Network with evidence from Case
#3 entered. Cases #3 and #4 show the same observed values but the domain
expert disagreed with the results of Case #4, for the above-mentioned reasons.
The following section summarizes the findings of this study.

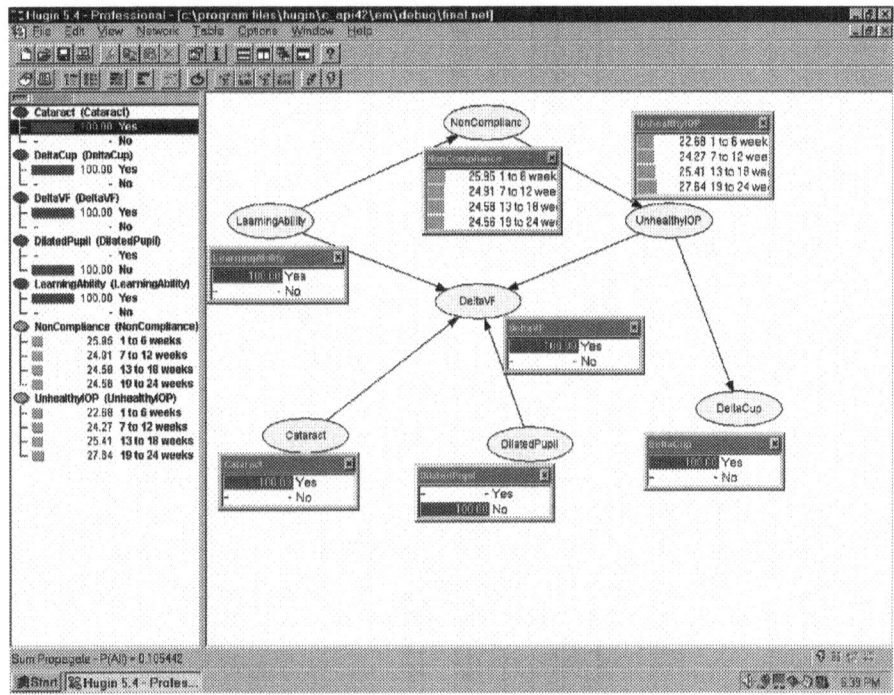

Figure 8. Bayesian Model showing evidence entered from Case #3

Findings

The findings of this research are presented along with the original objectives of this research. The objectives of this study were fulfilled to an acceptable degree of satisfaction. An outline of the main objectives and the ways in which this study achieved them, are briefly stated to put this research effort in its proper perspective:

1. *Identify the potential variables affecting the medication compliance process in the glaucoma domain.*

   In this research, the first step taken was to identify the variables affecting noncompliance among the glaucoma patients in consultation with a domain expert. Based on this expertise, the probabilities of each of the variables (prior or known) were ascertained. The patient data used in this study confirms that Dilated Pupil, Visual Field and Optic nerve Cup,

Cataract, and the comprehensive ability of the patient are good predictors of medication compliance in glaucoma cases.

Learning ability has a direct influence on the patient's compliance and also on Visual Field reading. Other variables that can affect Visual Field reading are Cataract and Dilated Pupil. When the patient has not complied with the medication, unhealthy intraocular pressure builds up causing the visual field loss and optic nerve cup damage.

2. *Model the degree of uncertainty involved in the medication compliance process.*

    The certainties associated with the identified variables were recognized early and were defined as complete data for some variables and as incomplete (or missing) data for others. Such uncertainty is typical and is an inherent problem in this type of research. The Bayesian Belief Network, which was employed in this research, is the methodology of choice and is ideally suited for empirical estimation, given the evidence. The interactions among the variables were modeled using cause-effect relationships. These relationships were quantified using probabilities.

    Thus the glaucoma noncompliance process was modeled using Bayesian Belief Network using qualitative structure (cause-effect relationships) and quantitative data (probabilities). Posterior probabilities were derived from the available patient data set. This was done using a Bayesian learning algorithm. The Bayesian modeling of glaucoma noncompliance process was successful and the results are very close in agreement to the domain expert's evaluation of patient cases.

3. *Capture the domain expert's experience in terms of subjective probabilities.*

    The ophthalmologist (domain expert) provided the prior probabilities based on his experience. The ophthalmologist's expertise comes from medical training, licensure, experience, and continuing education. The prior or known probabilities are a subjective interpretation of the identified variables by the domain expert. In the Bayesian model, a prior probability describes the strength of the belief, which a domain expert can justifiably hold that a certain statement of fact (relationship between variables) is indeed true. The prior probability values were refined by the clinical data used in this research.

4. *Identify the relevant techniques that are useful in reasoning in this domain.*

In this research, a Bayesian Belief Network and EM algorithm were identified as appropriate tools for determining noncompliance among glaucoma patients. These tools can be used in modeling domain problems when data is missing and when the relationships among the variables are acyclic in nature and have cause-effect relationships.

The cause-and-effect relations are modeled with probabilities. Bayesian framework uses a Bayesian probability theory and models the dependencies from structural descriptions of the problem. It was confirmed that Bayesian Networks offer a powerful framework to solve problems in domains where dependencies among variables are known.

Summary of Results

The Bayesian framework can be used to illustrate the complex nature of the glaucoma noncompliance process. Available patient cases can be used to update prior beliefs and to predict the outcome of future cases. In addition, the Bayesian model can be used to confirm clinical intuition or to formulate a prescription strategy for the patient. The outcome of this study supports agreement with the domain expert in three out of four cases. The results of this research showed that a detailed and accurate diagnosis with Bayesian framework was possible.

Results from this study could potentially improve the decision-making process, given the uncertain and incomplete data available to a physician. The Bayesian model may be generalized to other business situations where a decision has to be made based on incomplete or missing dataset.

Strategies for further improvement and enhancement of this model and goals for further research are as follows:

1. Further refinement of the model could include ranges for the values, rather than "*yes*" or "*no*", for even better prediction results.

2. A weakness in the design of the research was the limited availability of dataset to validate the model. A larger dataset could refine the model even better.

3. Only seven variables were used in order to limit the complexity of the glaucoma non-compliance process and to define the process. A weak-

ness in the model could be that simplifying assumptions were made in favor of defining the process.

4. The model could be expanded to include variables such as frequency of missed appointments, number of drugs prescribed, duration of drug intake, an unstable home or family situation, dissatisfaction with the treatment, and level of understanding of the disease and its treatment (Glaucoma Assoc. of New York, 2000).

5

Conclusions

This chapter presents the conclusions, implications of this research for the medical domain in general, and glaucoma treatment in particular, recommendations for future studies and a final summary. The goal of this research was to build a model-based decision support system using Bayesian Belief Networks (BBNs) to determine the compliance of glaucoma patients with prescribed medications. Specifically, this research aimed to achieve the following:

1. Identify potential variables affecting patient compliance behavior in glaucoma cases and investigate the inter-relationships among such variables.

2. Model the patient compliance problem and the strategies used by ophthalmologists as a BBN.

3. Estimate subjective probabilities to represent the interdependencies among the variables of interest in the BBN.

4. Use an appropriate algorithm to infer posterior probabilities for events of interest given a set of evidence.

5. Test and refine the BBN to verify the appropriateness of the model to predict compliance behavior with reasonable accuracy.

The rest of the chapter is organized as follows:

- Conclusions: This section describes a review of the results obtained through this research and compares them with the goals that had been identified for the research.

- Implications: This section discusses the potential impact of this research and how it might contribute to the medical domain as well as other areas of science and technology.

- Recommendations: This section provides the various possibilities for additional research that could be conducted to further enhance the modeling of the glaucoma non-compliance process.

- Summary: This section presents an overview of the entire research effort including goals, methodologies, specific procedures employed, results and implications.

Conclusions

The aim of this research was to identify the factors affecting noncompliance among glaucoma patients, define the relationships among them, and develop a suitable methodology to model their behavior. The modeling was successful and very close in agreement to that of the domain expert. Further, the Bayesian Belief Network was found to be suited for modeling the problem due to the cause-effect relationships among the variables and the acyclic nature of the relationships.

Domain knowledge acquired from ophthalmologists was used to identify the variables of interest. The variables were represented as nodes and the interactions among them were represented as directed arcs in the model. The dependencies between the variables were identified as causal arrows pointing from causes to effects. The direction of arc indicated the degree of influence or dependency between the variables. The available patient data confirms that Dilated Pupil, Visual Field and Optic nerve Cup, Cataract, and comprehensive ability of the patient are good predictors of medication compliance in glaucoma cases.

The prior probabilities (estimates of the known variables) from the domain expert were used in this research and further refined using the model. The prior probabilities provided a quantitative insight into the nature of the problem at hand and the inter-relationships among the variables.

In this research, the unknown variables in the model were estimated using Expectation–Maximization (EM) algorithm. The EM algorithm further refined the probabilities of the known variables, which are the posterior probabilities. The prior probabilities of the known variables were based on observation and the domain expert's estimates of the variables affecting the noncompliance among glaucoma patients.

The model was developed using a Bayesian Belief Network and refined using EM algorithm. The model was tested using patient datasets to verify its appropriateness to predict noncompliance behavior among glaucoma patients with reasonable accuracy.

The study included a restricted number of variables to identify a simple methodology using Bayesian Belief Networks to ascertain the approximate noncompliance and uncontrolled IOP. The included variables in their simplicity may limit the complexity that surrounds both the onset and the manifestation of glaucoma in affected patients. Thus, even with the recognized limitations, this research provided a good starting point and the potential to positively contribute to the existing literature and expand the relevant research horizon.

We have constructed a Bayesian model that consists of three components: a qualitative model, probabilities, and an inference system. The primary purpose of a Bayesian model is to give probabilistic estimates for events that are not directly observable. This model is an explicit representation of patient compliance problem in the glaucoma domain. The model was refined using clinical data and validated by comparing the predictions of the model with the performance of the domain expert.

The model can be refined to include other variables related to the glaucoma compliance process. A clinical decision support system can be developed using the refined model that can be used in preventing medical errors in assessing glaucoma compliance process. The results of this research potentially improve the decision-making process given the uncertain and incomplete data available to a physician. The model may be generalized to other business situations where a decision has to be made based on an incomplete and uncertain dataset.

Thus it can be seen, all the objectives of this research were fulfilled. The implications of this research and recommendations for future research are presented in the subsequent sections.

Implications

It is evident from this research that the Bayesian framework is an appropriate tool to model the uncertainties involved in the glaucoma noncompliance process. This research reviewed, incorporated, and made an effort to improve upon the known models, which can be potentially used to assess medication compliance among patients receiving medical treatment for glaucoma. The model provided approximate estimates of the duration of noncompliance and elevated IOP. This was beneficial for understanding the rate at which the disease worsens due to either noncompliance of medication, duration of elevated IOP, or both.

The model was found to be appropriate in predicting patient noncompliance and the duration of the elevated IOP among glaucoma patients with reasonable accuracy and can be used with appropriate modifications and refinements for other diseases characterized by inherent uncertainties.

Recommendations

As mentioned, this research is a first step in developing a Bayesian Belief Network and in identifying the factors affecting noncompliance among glaucoma patients. As with every initial research effort, this research has limitations and recognizes that more can be done in the future, anchored with our findings. The recommendations for future research are as follows:

1. A user interface can be developed with Hugin API to automate the derivation of posterior probabilities.

2. The results of this research were based on a limited patient dataset. A larger dataset may be sought to enhance the predictive value of the research effort and the attendant generalizations.

3. In the present study, we used four durations (*"1 to 6 weeks"*, *"7 to 12 weeks"*, *"13 to 18 weeks"*, and *"19–24 weeks"*) for the unhealthy IOP and non-compliance variables. These variables can be refined with experimental data and appropriate statistical analyses.

4. In this study, we used two states (*"changed"* or *"not changed"*) for DeltaVF and DeltaCup. The values of these variables can be further refined to include several states such as *"10% changed"*, *"20% changed"*

and so on. This would enhance the predictive ability of the model in those cases where two patients have similar states (*"Changed"*) but the variance of change is different.

5. The glaucoma model presented in this study included a restricted number of variables to identify a simple methodology using Bayesian Belief Networks to ascertain the approximate noncompliance and uncontrolled IOP. More variables could be included to ascertain their significance in determining the noncompliance process. According to the Glaucoma Associates of New York (2000), several additional variables can influence the glaucoma non-compliance process. These variables are the frequency of missed appointments, number of drugs prescribed, duration of drug intake, an unstable home or family situation, dissatisfaction with the treatment, or a poor understanding of the disease and its treatment. These variables could refine the complexity that surrounds both the onset and the manifestation of glaucoma in affected patients.

6. Another interesting area of research is to explore and evaluate this Bayesian model to determine noncompliance for other diseases, with appropriate modifications.

Summary

The purpose of this research was to devise a Bayesian framework to assess compliance with medication in glaucoma patients. This research applied Bayesian modeling to medication noncompliance in glaucoma. Glaucoma is a clinical condition characterized by increased intraocular pressure (IOP) in the eye that results in damage to the optic nerve, loss of visual field, and eventual blindness. It predominantly affects people over the age of sixty. IOP is controlled with several medications. As glaucoma disease affects mostly the elderly, compliance (usage of drugs at scheduled intervals as prescribed by a physician) is a major problem (Patel and Spaeth, 1995). Bayesian Networks have increasingly become the tools of choice in solving problems involving uncertainty in the medical domain. These models have been successfully applied to various medical diagnosis applications.

Compliance with medication is very important in the case of glaucoma because of its chronic nature. Compliance encompasses frequent medication along with awareness and interest in health, keeping appointments, and making appropriate lifestyle changes. Noncompliance can range from accidental lapses (missing medication at random) to premature cessation, or ineffective and incorrect methods of taking medications (Murphy and Coster, 1997; Rotchford and Murphy, 1998). If the patient is non-compliant, irrespective of the advances in medical field, the health of the patient will not improve (Bloch, Rosenthal, Friedman, and Caldarolla, 1997). Medication noncompliance behavior was analyzed and assessed in terms of cause-and-effect relationships using a model. Factors that affected compliance with medication and data that indicated the worsening of glaucoma were juxtaposed to identify the relationships among the variables.

The glaucoma model was developed by acquiring a qualitative model from an ophthalmologist (domain expert) for the compliance process in glaucoma patients. The qualitative model consisted of variables, which represent entities, and directed arcs that connected variables. The model was built in causal form to imply connections in the network pointing from causes to effects. The direction of arc indicated that there was some degree of influence or dependency between the variables. The Bayesian Network represented seven variables interacting with each other in the glaucoma domain. Learning ability had a direct influence on patient compliance and also on the visual field reading. Other variables that can affect visual field reading are Cataract and Dilated Pupil. When the patient had not complied with the medication, unhealthy intraocular pressure built up causing visual field loss and optic nerve cup damage. This model used a directed acyclic graph (DAG) that showed no cyclic path from any node.

Hugin Lite was used to develop the probability model and to refine the probabilities. The Hugin system is a popular tool for constructing model-based decision support systems in domains characterized by inherent uncertainty. The model supported Bayesian Belief Networks and influenced diagrams. Also, the Hugin tool consisted of an easy to use graphical user interface with an Application Program Interface (API) for integrating Hugin into any applications written in C or C++.

The prior probabilities were provided by the ophthalmologist (domain expert). In the Bayesian model, a probability describes the strength of the belief (prior probabilities) by which a domain expert can justifiably hold a certain statement-of-fact to be indeed, true. The posterior probabilities were derived from the available patient dataset and Expectation-Maximization (EM) algorithms were used to derive probabilities of missing variables.

The present research relied upon "criterion validity" which compares "the degree to which information provided by an instrument agrees with information obtained from other sources" (Fraenkel and Wallen, 1999). In this instance, the Bayesian model was used as the "criterion instrument." To assess research validity, the results obtained through the model were compared against the domain expert's evaluation of the patient cases. Since the results provided by the instrument agreed with the information provided by the domain expert, the Bayesian framework was considered reasonably valid.

The study included a restricted number of variables to identify a simple methodology using Bayesian Belief Networks to ascertain approximate non-compliance and uncontrolled IOP. The included variables in their simplicity may limit the complexity that surrounds both the onset and the manifestation of glaucoma in affected patients. Thus, even with the recognized limitations, this research provided a good starting point and the potential to positively contribute to the existing literature and expand the relevant research horizon.

The model can be refined to include other variables in the glaucoma compliance process. A clinical decision support system can be developed using the refined model and can be used in preventing medical errors in the glaucoma compliance process. The gains of this research potentially improve the decision-making process, given the uncertain and incomplete data available to a physician. The model may be generalized to other business situations where a decision has to be made based on incomplete and uncertain datasets.

References

Andreassen, S., Woldbye, M., Falk, B., and Andersen, S. K. (1987). MUNIN–A causal probabilistic network for interpretation of electromyographic findings. <u>Proceedings of the Tenth International Joint Conference on Artificial Intelligence</u>, 366-372.

Andreassen, S., Hovorka, R., Benn, J., Olesen, K., and Carson, E. (1991). A model-based approach to insulin adjustment. <u>Proceedings of 3<u>rd</u> conference on Uncertainty in Artificial Intelligence,</u> 239-248.

Andreassen, S. (1994). Model-based bio-signal interpretation. <u>Methods of Information in Medicine, 33,</u> 103-110.

Berzuini, C., Bellazi, R., Quaglini, S., and Spiegelhalter, D. J. (1992). Bayesian networks for patient monitoring. <u>Artificial Intelligence in Medicine, 4,</u> 243-260.

Bellazi, R., Berzuini, C., Quaglini, S., Spiegelhalter, D. J., and Leaning, M. (1991). Cytotoxic chemotherapy monitoring using stochastic simulation on graphical models. In M. Stefanelli, A. Hasman, M. Fieschi and J. Talmon (Eds.), <u>Proceedings of the Third Conference on Artificial Intelligence in Medicine, Lecture Notes in Medical Informatics 44,</u> (pp. 227-238) New York, NY: Springer Verlag Publishers.

Bellazzi. R. and Riva. A (1995). Learning conditional probabilities with longitudinal data, In <u>IJCAI, 7-15</u>. Montreal, Quebec: Canada, August 21-25, 1995.

Bellazzi. R. and Riva. A (1998, September). Learning bayesian networks probabilities with longitudinal data. IEEE Transactions on Systems, Man, and Cybernetics-Part A: Systems and Humans, 28 (5) 629-636.

Binder, J., Koller, D., Russell, S., and Kanazawa, K. (1997). Adaptive probabilistic networks with hidden variables. Machine Learning, 29, 213-214.

Bloch, S., Rosenthal, A. R., Friedman, L., and Caldarolla, P. (1997, August). Patient compliance in glaucoma. British Journal of Ophthalmology, 61, 531-534.

Brown, M. M., Brown, G. C., and Spaeth, G. L. (1984, February). Improper topical self-administration of ocular medication among patients with glaucoma. Canadian Journal of Ophthalmology, 19, 2-5.

Bunn, C. C., Du, M., Niu, K., Johnson, T. R., Poston, W. S. C., and Foreyt, J. P. (1999). Predicting the risk of obesity using a bayesian network. Proceedings of American Medical Informatics Association Conference, Booth number 64, 1999.

Chang, J. S., Lee, D. A., Petursson, G., Spaeth, G., Zimmerman T. J., Hoskins, H. D., Mills, R., Brown, R., Kass, M., and Lue, J. (1991, Summer). The effect of a glaucoma medication reminder cap on patient compliance and intraocular pressure. Journal of Ocular Pharmacology, 7, 117-124.

Chevrolat, J., Golmard, J., Ammar, S., Jouvent, R., and Boisvieux, J. (1998). Modelling behavioral syndromes using Bayesian networks. Artificial Intelligence Medicine, 14, 259-277.

Cooper, G. F., and Herskovits, E. (1992). A bayesian method for the induction of probabilistic networks from data. Machine Learning, 9, 309-347.

Cooper, J. (1996). Improving compliance with glaucoma eye-drop treatment. Nursing Times, 92 (32), 7-13.

Cramer, J. A. (1998). Enhancing patient compliance in the elderly: Role of packaging aids and monitoring. Drug and Aging, 12 (1), 7-15.

Dempster, A. P., Laird, N. M., and Rubin, D. B. (1977). Maximum likelihood from incomplete data via the EM algorithm. Journal of the Royal Statistical Society, B39, 1-38.

de Dombal, F., Leaper, D., Staniland, J., Horrocks, J., McCann, A. (1972). Computer aided diagnosis of acute abdominal pain. British Medical Journal, 2, 9-13.

Duane, T.D., and Jaeger, E.A. (1988). Clinical Ophthalmology, Vol. 3, PA: J. B. Lippincott Company.

Fraenkel, J and Wallen, N. (1999). How to design and evaluate research in education. New York, NY: McGraw-Hill Publishing.

Friedman, N. (1997). Learning Bayesian networks in the presence of missing values and hidden variables. In D.Fisher (Ed.), Proceedings of the Fourteenth International conference on Machine Learning. San Francisco, CA: Morgan Kaufmann.

Friedman, N. (1998). The Bayesian structural EM algorithm. In G.F.Cooper and S.Moral (Eds.), Proceedings of Fourteenth Conference on Uncertainty in Artificial Intelligence (UAI'98). San Francisco, CA: Morgan Kaufmann.

Furness, P. N., Kazi, J. I., Nicholson, M. L., Kirkpatrick, U., Taub, N., Davies, D., and Solez, K., (1997). The UK assessment of the Banff classification of transplant pathology and a neural network approach to improved diagnosis of acute rejection. Unpublished manuscript.

Gaebel, W. (1997). Towards the improvement of compliance: The significance of psycho-educational and new anti-psychotic drugs. International Clinical of Psychopharmacology, 12 (Suppl. 1:S), 37-42.

Geman, S and Geman, D. (1984). Stochastic relaxation, Gibbs distributions, and the Bayesian restoration of images. IEEE Transactions in Pattern Analysis and Machine Intelligence, PAMI, 6 (9), 721-741.

Glaucoma Associates of New York (2000). Compliance: The patient's end of the bargain. [On-line] Available: <http://www.glaucoma.net>

Gorry, G. (1973). Computer-assisted clinical decision making, <u>Methods of Information in Medicine, 12,</u> 45-51.

Jensen, F. (1997). <u>Introduction to Bayesian networks.</u> New York, NY: Springer-Verlag Publishing.

Heckerman, D., and Nathwani, B. (1992). Toward normative expert systems: Part I—the Pathfinder project. <u>Methods of Information in Medicine, 31,</u> 90-105.

Heckerman, D., Horvitz, E., and Nathwani, B. (1992). Toward normative expert systems: Part II—probability-based representations for efficient knowledge acquisition and inference. <u>Methods of Information in Medicine, 31,</u> 106-116.

Hugin. (2000). HUGIN API reference manual version 4.2. [On-line] Available: <http://www.hugin.dk/>

Kahn, C., Roberts, L., Wang, K., Jenks, D., Haddawy, P. (1995). Preliminary investigation of a Bayesian network for mammographic diagnosis of breast cancer. <u>Proceedings of the 19th Annual Symposium on Computer Applications in Medical Care</u> (pp. 208-212). Philadelphia, PA: Hanley and Belfus.

Korver, M., and Lucas. P. (1993). Converting a rule-based expert system into a belief network. <u>Medical Informatics, 18</u> (3), 219-241.

Kosoko, O., Quigley, H. A., Vitale S., Enger, C., Kerrigan, L., and Tielsch, J. M. (1998). Risk factors for non-compliance with glaucoma follow-up visits in a resident's eye clinic. <u>Ophthalmology, 105</u> (11), 2105-11.

Landrum, M., and Normand, S.(1999). Applying Bayesian ideas to the development of medical guidelines. <u>Statistics Medicine, 18,</u> 117-137.

Larizza, C, Bellazzi. R, Riva, A. (1997). Temporal abstractions for diabetic patients management. In E. Keravnou, C. Garbay, R. Baud, and J. Wyatt (Eds.) <u>Proceedings of AIME 97, number 1211 in Lecture Notes in Computer Science</u> (pp. 319-330). New York, NY: Springer-Verlag Publishing.

Lauritzen, S. (1995). The EM algorithm for graphical association models with missing data. Computational Statistics and Data Analysis, 19, 191-201.

Little, R. J. A. and Rubin, D. B. (1987). Statistical analysis with missing data. New York: Wiley Publishing.

Little, R. and Schenker, N. (1995). Missing data. In G. Arminger, C. Clogg, & M. Sobel (Eds.), Handbook of Statistical Modeling for the Social and Behavioral Sciences (pp. 39-75). New York: Plenum.

Mani, S., McDermott, S., and Valtorta., M (1997). MENTOR: A Bayesian model for prediction of mental retardation in newborns. Journal of Research in Developmental Disabilities, 18 (5), 303-318.

Melia, M. and Jordan, M. I. (1998). Estimating dependency structure as a hidden variable. In Advances in Neural Information Processing Systems, 10, Jordan, M. I., Kearns, M. J. and Solla, S. A. (Eds.), Cambridge, MA: MIT Press.

Mezzetti, M. and Robertson, C. (1999). A hierarchical Bayesian approach to age-specific back-calculation of cancer incidence rates. Statistics Medicine, 18, 919-933.

Miller, M. and Seaman, J. (1998). A Bayesian approach to assessing the superiority of a dose combination. Biometrical Journal, 40, 43-55.

Murphy, J. and Coster G. (1997). Issues in patient compliance [Review of the book Issues in Patient Compliance]. Department of Medicine, University of Auckland, School of Medicine, New Zealand, 54 (6), 797-800.

Oppel, U. G., Hierle, A., Janke, L., and Moser, W. (1993). Transformations of compartmental models into sequences of causal probabilistic networks. Artificial Intelligence in Medicine Journal, 319-330.

Patel, S. C. and Spaeth, G. L. (1995). Compliance in patients prescribed eyedrops for glaucoma. Ophthalmic Surgery, 26 (3), 233-236.

Quaglini, S., Bellazzi, R., Stefanelli, M., and Locatelli, F. (1993). Sharing and reusing therapeutic knowledge for managing leukemic children. In S.

Andreassen, R. Engelbrecht, J. Wyatt (Eds.), AIME93: Proceedings of 4th Conference on Artificial Intelligence in Medicine (pp. 319-330). Amsterdam: ISOP Press.

Riva. A and Bellazzi. R (1996). Learning temporal probabilistic causal models from longitudinal data. Artificial Intelligence in Medicine Journal, 8, 217-234.

Rotchford, A. P. and Murphy, K. M. (1998). Compliance with timolol treatment in glaucoma. Eye, 12 (2), 234-236.

Rubin, D.B. (1987). Multiple imputation for nonresponse in surveys. New York: John Wiley & Sons Publishing.

Singh, M. (1997). Learning Bayesian networks from incomplete data. Proceedings of the 14th National Conference on Artificial Intelligence (AAAI'97), Providence, RI, July 27–31.

Singh, M. (1998, May). Learning Bayesian Networks for Solving Real-World Problems. Unpublished doctoral dissertation, Computer and Information Science, University of Pennsylvania, 1998.

Spiegelhalter, D. J., Dawid P. A., Lauritzen, L. L, and Cowell, R. G. (1993). Bayesian analysis in expert systems. Statistical Science, 219-283.

Thiesson, B. (1995). Accelerated quantification of Bayesian networks with incomplete data. Proceedings of the First International Conference on Knowledge Discovery and Data Mining (KDD-95), (pp.306-311). Montreal, Canada: AAA Press.

Thiesson, B., Meek, C., Chickering, M., and Heckerman, D. (1998). Learning mixtures of Bayesian networks. In G.F. Cooper and S.Moral (Eds.), Proceedings of the Fourteenth Conference on Uncertainty in Artificial Intelligence (UAI'98). San Francisco, CA: Morgan Kaufmann.

Wiegerinck, W., Kappen, H., Neijt, J., Van Dam, P., Braak, W., and Burg, W., (1999). Inference and advisory system for medical diagnosis. [On-line] Available: <http://www.mbfys.kun.nl/SNN/Research/node37.html>

APPENDIX A

Actual.cpn File

abcd "abcd"   % Domain name and label

100 40   % node dimensions (pt)

1   % scale factor

% Node no. 1 Learning Ability

N1 "1" 0 230

("1" "2")

(0.85 0.15)

% Node no. 2 Cataract

N2 "2" 10 60

("1" "2")

(0.6 0.4)

% Node no. 3 Dilated Pupil

N3 "3" 160 0

("1" "2")

(0.4 0.6)

% Node no. 4 Non-Compliance

N4 "4" 160 280

("1" "2" "3" "4") N1

((0.5 0.25 0.15 0.1)

 (0.03 0.07 0.3 0.6))

% Node no. 5 Unhealthy IOP

N5 "5" 330 220

("1" "2" "3" "4") N4

((0.9 0.05 0.03 0.02)

 (0.05 0.9 0.02 0.03)

 (0.03 0.02 0.9 0.05)

 (0.02 0.03 0.05 0.9))

% Node no. 6 DeltaCup

N6 "6" 400 30

("1" "2") N5

((0.2 0.8)   % 1 to 6 weeks value 1

 (0.5 0.5)   % 7 to 12 weeks value 2

 (0.7 0.3)   % 13 to 18 weeks value 3

 (0.95 0.05))   % 19 to 24 weeks value 4

% Node no. 7 Delta VF

N7 "7" 180 140

("1" "2") N1 N2 N3 N5

(((((0.45 0.55)	% Yes Yes Yes 1 to 6 weeks
(0.6 0.4)	% Yes Yes Yes 7 to 12 weeks
(0.85 0.15)	% Yes Yes Yes 13 to 18 weeks
(0.95 0.05))	% Yes Yes Yes 19 to 24 weeks
((0.5 0.5)	% Yes Yes No 1 to 6 weeks
(0.65 0.35)	% Yes Yes No 7 to 12 weeks
(0.9 0.1)	% Yes Yes No 13 to 18 weeks
(0.95 0.05)))	% Yes Yes No 19 to 24 weeks
(((0.35 0.65)	% Yes No Yes 1 to 6 weeks
(0.5 0.5)	% Yes No Yes 7 to 12 weeks
(0.75 0.25)	% Yes No Yes 13 to 18 weeks
(0.95 0.05))	% Yes No Yes 19 to 24 weeks
((0.4 0.6)	% Yes No No 1 to 6 weeks
(0.55 0.45)	% Yes No No 7 to 12 weeks
(0.8 0.2)	% Yes No No 13 to 18 weeks
(0.95 0.05))))	% Yes No No 19 to 24 weeks
((((0.25 0.75)	% No Yes Yes 1 to 6 weeks
(0.4 0.6)	% No Yes Yes 7 to 12 weeks
(0.7 0.3)	% No Yes Yes 13 to 18 weeks
(0.9 0.1))	% No Yes Yes 19 to 24 weeks

((0.3 0.7)	% No Yes No 1 to 6 weeks
(0.45 0.55)	% No Yes No 7 to 12 weeks
(0.75 0.25)	% No Yes No 13 to 18 weeks
(0.95 0.05)))	% No Yes No 19 to 24 weeks
(((0.15 0.85)	% No No Yes 1 to 6 weeks
(0.3 0.7)	% No No Yes 7 to 12 weeks
(0.6 0.4)	% No No Yes 13 to 18 weeks
(0.8 0.2))	% No No Yes 19 to 24 weeks
((0.2 0.8)	% No No No 1 to 6 weeks
(0.35 0.65)	% No No No 7 to 12 weeks
(0.65 0.35)	% No No No 13 to 18 weeks
(0.85 0.15)))))	% No No No 19 to 24 weeks

Glaucoma.dat File

7

2 2 2 4 4 2 2

15

0

1 2 2 0 0 1 1

1 2 2 0 0 1 1

2 2 2 0 0 1 1

2 2 2 0 0 1 1

2 1 1 0 0 1 1

1 1 1 0 0 1 2

1 1 1 0 0 2 2

1 1 1 0 0 2 2

1 1 1 0 0 2 1

1 2 1 0 0 1 1

1 1 2 0 0 1 1

1 2 2 0 0 1 1

1 1 2 0 0 1 1

1 1 2 0 0 1 1

1 2 2 0 0 1 1

APPENDIX B

Glaucoma.cpn File

abcd "abcd"   % Domain name and label

100 40   % node dimensions (pt)

1   % scale factor

% Node no. 1

N1 "1" 0 0

("1" "2")

(0.764706 0.235294) %

% Node no. 2

N2 "2" 100 0

("1" "2")

(0.529412 0.470588) %

% Node no. 3

N3 "3" 200 0

("1" "2")

(0.411765 0.588235) %

% Node no. 4

N4 "4" 300 0

("1" "2" "3" "4") N1

((0.259573 0.249143 0.245840 0.245444) % 1

(0.248469 0.248987 0.249864 0.252680)) % 2

% Node no. 5

N5 "5" 400 0

("1" "2" "3" "4") N4

((0.242976 0.246063 0.250950 0.260011) % 1

(0.239408 0.249947 0.250863 0.259782) % 2

(0.239196 0.245879 0.255314 0.259611) % 3

(0.237916 0.244672 0.249561 0.267851)) % 4

% Node no. 6

N6 "6" 0 100

("1" "2") N5

((0.670741 0.329259) % 1

(0.688772 0.311228) % 2

(0.700192 0.299808) % 3

(0.719842 0.280158)) % 4

% Node no. 7

N7 "7" 100 100

("1" "2") N1 N2 N3 N5

(((((0.419878 0.580122) % 1 1 1 1

(0.418000 0.582000) % 1 1 1 2

(0.416292 0.583708) % 1 1 1 3

(0.412470 0.587530)) % 1 1 1 4

((0.623960 0.376040) % 1 1 2 1

(0.632547 0.367453) % 1 1 2 2

(0.638468 0.361532) % 1 1 2 3

(0.649524 0.350476))) % 1 1 2 4

(((0.550090 0.449910) % 1 2 1 1

(0.553861 0.446139) % 1 2 1 2

(0.556513 0.443487) % 1 2 1 3

(0.561600 0.438400)) % 1 2 1 4

((0.652393 0.347607) % 1 2 2 1

(0.662315 0.337685) % 1 2 2 2

(0.669077 0.330923) % 1 2 2 3

(0.681528 0.318472)))) % 1 2 2 4

(((((0.550044 0.449956) % 2 1 1 1

(0.553844 0.446156) % 2 1 1 2

(0.556514 0.443486) % 2 1 1 3

(0.561659 0.438341)) % 2 1 1 4

((0.500000 0.500000) % 2 1 2 1

(0.500000 0.500000) % 2 1 2 2

(0.500000 0.500000) % 2 1 2 3

(0.500000 0.500000))) % 2 1 2 4

(((0.500000 0.500000) % 2 2 1 1

(0.500000 0.500000) % 2 2 1 2

(0.500000 0.500000) % 2 2 1 3

(0.500000 0.500000)) % 2 2 1 4

((0.590345 0.409655) % 2 2 2 1

(0.596998 0.403002) % 2 2 2 2

(0.601636 0.398364) % 2 2 2 3

(0.610481 0.389519))))) % 2 2 2 4

Glaucoma.out File

Likelihood for iteration no 0 = -64.898166

Iteration No. 1

Likelihood for iteration no 1 = -62.685486

Iteration No. 2

Likelihood for iteration no 2 = -62.831100

Iteration No. 3

Likelihood for iteration no 3 = -62.878433

Iteration No. 4

Likelihood for iteration no 4 = -62.898202

Iteration No. 5

Likelihood for iteration no 5 = -62.908501

Iteration No. 6

Likelihood for iteration no 6 = -62.914406

Iteration No. 7

Iteration 7 converges; returning cpn_file generated in iteration 7

EM likelihood = -62.914406

Final.net File

abcd "abcd"   % Domain name and label

100 40   % node dimensions (pt)

1   % scale factor

% Node no. 1 Learning Ability

LearningAbility "LearningAbility" 0 230

("Yes" "No")

(0.764706 0.235294) %

% Node no. 2 Cataract

Cataract "Cataract" 10 60

("Yes" "No")

(0.529412 0.470588) %

% Node no. 3 Dilated Pupil

DilatedPupil "DilatedPupil" 160 0

("Yes" "No")

(0.411765 0.588235) %

% Node no. 4 Non-Compliance

NonCompliance "NonCompliance" 160 280

("1 to 6 weeks" "7 to 12 weeks" "13 to 18 weeks" "19 to 24 weeks") Learning-
Ability

((0.259573 0.249143 0.245840 0.245444) % 1

(0.248469 0.248987 0.249864 0.252680)) % 2

% Node no. 5 Unhealthy IOP

UnhealthyIOP "UnhealthyIOP" 330 220

("1 to 6 weeks" "7 to 12 weeks" "13 to 18 weeks" "19 to 24 weeks") Non-Compliance

((0.242976 0.246063 0.250950 0.260011) % 1

(0.239408 0.249947 0.250863 0.259782) % 2

(0.239196 0.245879 0.255314 0.259611) % 3

(0.237916 0.244672 0.249561 0.267851)) % 4

% Node no. 6 DeltaCup

DeltaCup "DeltaCup" 400 30

("Yes" "No") UnhealthyIOP

((0.670741 0.329259) % 1

(0.688772 0.311228) % 2

(0.700192 0.299808) % 3

(0.719842 0.280158)) % 4

% Node no. 7 Delta VF

DeltaVF "DeltaVF" 180 140

("Yes" "No") LearningAbility Cataract DilatedPupil UnhealthyIOP

(((((0.419878 0.580122) % 1 1 1 1

(0.418000 0.582000) % 1 1 1 2

(0.416292 0.583708) % 1 1 1 3

(0.412470 0.587530)) % 1 1 1 4

((0.623960 0.376040) % 1 1 2 1

(0.632547 0.367453) % 1 1 2 2

(0.638468 0.361532) % 1 1 2 3

(0.649524 0.350476))) % 1 1 2 4

((((0.550090 0.449910) % 1 2 1 1

(0.553861 0.446139) % 1 2 1 2

(0.556513 0.443487) % 1 2 1 3

(0.561600 0.438400)) % 1 2 1 4

(((0.652393 0.347607) % 1 2 2 1

(0.662315 0.337685) % 1 2 2 2

(0.669077 0.330923) % 1 2 2 3

(0.681528 0.318472)))) % 1 2 2 4

(((((0.550044 0.449956) % 2 1 1 1

(0.553844 0.446156) % 2 1 1 2

(0.556514 0.443486) % 2 1 1 3

(0.561659 0.438341)) % 2 1 1 4

(((0.500000 0.500000) % 2 1 2 1

(0.500000 0.500000) % 2 1 2 2

(0.500000 0.500000) % 2 1 2 3

(0.500000 0.500000))) % 2 1 2 4

((((0.500000 0.500000) % 2 2 1 1

(0.500000 0.500000) % 2 2 1 2

(0.500000 0.500000) % 2 2 1 3

(0.500000 0.500000)) % 2 2 1 4

(((0.590345 0.409655) % 2 2 2 1

(0.596998 0.403002) % 2 2 2 2

(0.601636 0.398364) % 2 2 2 3

(0.610481 0.389519))))) % 2 2 2 4

978-0-595-36839-6
0-595-36839-5